£18-75

CW01246259

Science and Convention

*Essays on Henri Poincaré's Philosophy of Science
and The Conventionalist Tradition*

FOUNDATIONS & PHILOSOPHY OF SCIENCE & TECHNOLOGY SERIES

General Editor: MARIO BUNGE, McGill University, Montreal, Canada

Some Titles in the Series

AGASSI, J.
The Philosophy of Technology

ALCOCK, J.
Parapsychology: Science or Magic?

ANGEL, R.
Relativity: The Theory and its Philosophy

BUNGE, M.
The Mind-Body Problem

HATCHER, W.
The Logical Foundations of Mathematics

SIMPSON, G.
Why and How: Some Problems and Methods in Historical Biology

WILDER, R. L.
Mathematics as a Cultural System

Pergamon Journals of Related Interest

STUDIES IN HISTORY AND PHILOSOPHY OF SCIENCE*
Editor:
Prof. Gerd Buchdahl, Department of History and Philosophy of Science, University of Cambridge, England

This journal is designed to encourage complementary approaches to history of science and philosophy of science. Developments in history and philosophy of science have amply illustrated that philosophical discussion requires reference to its historical dimensions and relevant discussions of historical issues can obviously not proceed very far without consideration of critical problems in philosophy. *Studies* publishes detailed philosophical analyses of material in history of the philosophy of science, in methods of historiography and also in philosophy of science treated in developmental dimensions.

*Free specimen copies available on request

Science and Convention

*Essays on Henri Poincaré's Philosophy of Science
and The Conventionalist Tradition*

by

JERZY GIEDYMIN
University of Sussex, Brighton

PERGAMON PRESS
OXFORD · NEW YORK · TORONTO · SYDNEY · PARIS · FRANKFURT

U.K.	Pergamon Press Ltd., Headington Hill Hall, Oxford OX3 0BW, England
U.S.A.	Pergamon Press Inc., Maxwell House, Fairview Park, Elmsford, New York 10523, U.S.A.
CANADA	Pergamon Press Canada Ltd., Suite 104, 150 Consumers Road, Willowdale, Ontario M2J 1P9, Canada
AUSTRALIA	Pergamon Press (Aust.) Pty. Ltd., P.O. Box 544, Potts Point, N.S.W. 2011, Australia
FRANCE	Pergamon Press SARL, 24 rue des Ecoles, 75240 Paris, Cedex 05, France
FEDERAL REPUBLIC OF GERMANY	Pergamon Press GmbH, 6242 Kronberg-Taunus, Hammerweg 6, Federal Republic of Germany

Copyright © 1982 Jerzy S. Giedymin
All Rights Reserved. No part of this publication may be reproduced, stored in a retrieval system or transmitted in any form or by any means; electronic, electrostatic, magnetic tape, mechanical, photocopying, recording or otherwise, without permission in writing from the publishers.

First edition 1982

Library of Congress Catalog Card no: 81-81334

British Library Cataloguing in Publication Data

Giedymin, Jerzy
Science and convention. - (Foundations & philosophy of science & technology)
1. Science - Philosophy
I. Title
501   Q175

ISBN 0-08-025790-9

Printed in Great Britain by A. Wheaton & Co. Ltd., Exeter

*To the memory
of my parents*

Preface

ESSAYS in this volume are concerned with Henri Poincaré's philosophy of science—physics in particular—and with the conventionalist tradition in philosophy which he revived and reshaped at the turn of the century, simultaneously with, but independently of, Pierre Duhem. My aim was to make explicit the main ideas of that philosophy (Essays 1 and 5), to trace at least some of its historical background (Essays 1, 2 and 3) and to follow some of its later developments (Essays 2 and 4). Since the ideas in question are, for the most part, epistemological in nature and are discussed here in their historical context, the essays may perhaps be characterised as belonging to comparative epistemology.

The main reason for writing these essays was this. On the basis of my research I came to the conclusion that the philosophy of Poincaré and of his fellow-conventionalists had been perceived in too narrow a fashion, misunderstood and, consequently, underestimated. It is my belief that in Poincaré's philosophical writings one of the leading traditions in 19th century mathematics found its epistemological and ontological expression, viz. the tradition, represented in the writings of Galois, Poncelet, Plücker, Cayley, Hermite, Darboux, Klein and many others, for which the concepts of transformations, groups and invariants were fundamental and characteristic. It penetrated not only many branches of pure mathematics but also kinematics, analytic dynamics, optics and electrodynamics. It provided the foundation for the mathematical shape of much of the new 20th century physics of relativity and of the quanta. As regards Poincaré's own research, discontinuous groups were essential for the theory of automorphic functions (of one complex variable) which he discovered, stimulated by Hermite's work on modular functions; Lie's theory of transformation groups was the basis for his analysis of the foundations of geometry and—as Hadamard put it (1914, 1921, p. 220)—"in his guiding the nascent theory of relativity". Not surprisingly, the general ideas of transformations, groups and their structures as well as of a theory of invariants with respect to a group, which may be identified with the relativity theory of the group, became fundamental for Poincaré's epistemology (Essays 1, 2 and 5).

However, the outlined perception of Poincaré's philosophy is *not* the one shared by the majority of his commentators, nor is it reflected in the name "conventionalism" (geometrical or physical) now usually attached to it. For some reason, those whose pronouncements were to set the trend over roughly sixty years in defining and assessing Poincaré's contribution to the philosophy of physics, have restricted their attention to only one of its two interrelated features, viz. the one expressed in the claims such as that the axioms of geometry are conventions, that the choice of one of metric geometries is (empirically) arbitrary, that scientists often elevate empirical generalisations to the status of conventional principles, that some hypotheses (indifferent ones) are conventions freely invented by the mind, etc. This feature, undoubtedly important, became henceforth enshrined in the name conventionalism, never used by Poincaré himself. So, for example, as early as April 1922 Albert Einstein is reported as saying—at a discussion in Paris where he attended a reception in his honour—that there existed two opposite views concerning physical theories between which he felt unable to decide: Kant's apriorism and Poincarés conventionalism. Both were in agreement on the point that to construct science we need arbitrary concepts; but according to Kant these were given in *a priori* intuition whereas according to Poincaré they were conventions (*Bulletin de la Société Française de Philosophie*, Juillet 1922; reported in *Nature*, August 18, 1923, p. 253). Through a kind of "harmony of illusions" (to use L. Fleck's terms) this line of exegesis became firmly established in the German-speaking countries where, in all probability, the term "conventionalism" was coined. Roughly between 1918 and 1960 it permeated the logical empiricist perception of Poincaré's philosophy through the writings of Schlick, Reichenbach and Frank. It also dominated the view of the critics of logical empiricism under the influence of Popper, Frank and Holton, the first of whom saw instrumentalism (theories are nothing but mathematical contrivances) as an essential part of conventionalism. And yet, any attentive reader of *Science and Hypothesis*—not to mention other writings of Poincaré, especially the (1898) paper—should have noticed the other side of Poincaré's philosophy, the side concerned with the cognitive role of groups and their invariants in mathematics and physics. This can be illustrated by the following claims: the object of geometry is to study a particular group; in our minds the latent idea of a number of groups pre-exists (these are groups with which Lie's theory is concerned); from all possible groups we choose one to which to refer natural phenomena; there is no absolute space, there is no absolute time, we have no direct intuition of the simultaneity of distant events; the principle of relative motion is impressed upon us because the commonest experiments confirm it and because the consideration of the contrary hypothesis is singularly repugnant to the mind, etc. The concept of invariance (invariantism) or of relativity would be much more appropriate than "conventionalism", to refer to this feature of Poincaré's epistemology.

If now one inquires how the two aspects of Poincaré's epistemology, viz. conventionalism and invariantism came to be associated, then my answer is roughly this:

Poincaré's main epistemological problem was the question how objective knowledge and continuous progress were possible in spite of apparently disruptive changes in mathematics and in science. It is in his solution of this epistemological problem that he combined the ideas of convention and conventionality with the ideas of invariance, groups, transformations, etc.

From such developments as the discovery of non-Euclidean geometries and of higher spaces, from the criticism of the assumptions of absolute space and time in Newtonian mechanics and from the philosophy of the physics of the principles, Poincaré drew the conclusion that many elements of science which had appeared as ultimate truths about the world turned out, on closer analysis, to be conventional contrivances: though undoubtedly serving a purpose, they were substitutable by different conventions without change in the cognitive content of relevant theories. Consequently, not all problems and disputes in the history of science which had appeared as substantive and empirical were indeed such. What one did not see properly was that these disputes resulted from the existence of observationally equivalent (or, at least, experimentally indistinguishable) theories based on different conventions. Having discovered the non-empirical nature of some of the time-honoured problems concerned, for example, with the nature of space, time, light, etc., some philosophers, viz. nominalists, have concluded that "anything goes", that science is purely conventional. In fact it is such only if one adopts the nominalist attitude, otherwise it is not. However, in order to avoid the Scylla of naïve empiricism and the Charybdis of nominalism, one has to be clear about what constitutes the cognitive content of scientific theories and in effect where the limits of knowledge are. One way of doing this is to adopt the view, known as instrumentalism, according to which only the observational but not the abstract part of a scientific theory contributes to the cognitive (descriptive) content. This is Ramsey's view and, perhaps, Duhem's, but—despite all appearances—not Poincaré's. For, according to Poincaré, mathematics is not concerned with the production of purely formal games. It is a study of structures and mappings which, in its progress, reveals previously unsuspected connections between remote areas. Mathematical intuition plays the same active role in theoretical physics except that it is constrained there by the requirement of numerical (experimental) adequacy. The content of a theory consists, therefore, of its numerical predictions and of the form of its equations. The latter may be understood in terms of the group admitted (in Lie's sense) by the equations, and its invariants. Theoretical changes which do not alter either numerical predictions or the form of the theory are inessential. Even essential changes, however, need not be disruptive. Comparability is preserved if, for example, the groups of two apparently rival theories are shown to be subgroups of a third group; such is the case of classical and relativistic mechanics since the Galileo–Newtonian and Lorentzian groups are sub-groups of the affine group, as has been shown by F. Klein. In general, Poincaré's epistemology emphasises the relational nature of knowledge, whether in mathematics or in physics, and the unknowability of objects as such, i.e. apart from their relations to others. It is not equivalent to instrumentalism and might best be called *structural realism*. Since it combines elements of empiricism and rationalism it is a semi-rationalist view.

The outlined interpretation of Poincaré's epistemology has the advantage that it does justice to a lot of textual evidence ignored in the traditional interpretation which is in terms of conventionalism to the exclusion of invariantism or structuralism. Moreover, it has immediate implications for the debate over the discovery of special relativity and Poincaré's role in it. Some of the claims made in that debate, for example that owing to his alleged inductivism Poincaré was unable to conceive of a general relativity theory ("general", not in the sense of "general covariance" but in the sense of general principles applicable to more than one specific area of physics), must be seen at once as untenable, to say the least. Poincaré's epistemology was certainly much more rationalistic than the Machist epistemology of Einstein around 1905. In fact Einstein embraced a semi-rationalist philosophy, similar to Poincaré's, only in his later, post-Machist period. Since the most general concept of relativity theory, in the sense of an invariant theory of most comprehensive groups, was the basic element of the mathematical and philosophical tradition of which Poincaré was a creative member, he was naturally predisposed to apply such mathematical ideas in dealing with the problems in theoretical physics that arose from Michelson–Morley's and related experiments. On the question of Poincaré's role in the discovery of relativity theory, my position (Essay 5) differs both from the established view and Whittaker's controversial claims; it is as follows: the priority for the non-mathematical (experimental) formulation of the (special) relativity principle as well as for the statement of the limiting and constant velocity of light, are credited to Poincaré; the formulation of the relativity principle in terms of the Lorentz transformations and its mathematical elaboration are claimed to be a case of simultaneous and independent discovery by Poincaré and Einstein, with the latter seeing more clearly the physical implications and Poincaré being superior in the mathematics of the theory (invariants of the Lorentz group, basic ideas of the four-vector formalism, etc.).

In the light of what has been said already about the customary narrow interpretation of Poincaré's epistemology, it is not surprising that its historical background had been seen in a similarly restrictive fashion. Though Maxwell's and Lorentz's theories have figured prominently in the discussions of its origins, little else is usually mentioned and even the role of those two theories is often described in a slanted way. So, for example, Duhem, who was perhaps one of the first to emphasise the role of Maxwell's theory for the genesis of Poincaré's philosophy, saw this role exclusively in the alleged effect of Maxwell's instrumentalist view of (mechanical) models. Poincaré's analysis of the formal analogies between the equations of optics and electricity (in Poincaré's 1895) and his emphasis on their epistemological importance escaped Duhem's attention. Again, in recent contributions to the debate over the discovery of relativity theory, the impact of Lorentz's theory on Poincaré's world-view has been claimed (e.g. by Miller) with little regard for those features of Poincaré's philosophy which are clearly incompatible with Lorentz's realist interpretation of the ether hypothesis and the electromagnetic world view. I felt, therefore, that it was important to draw attention—however unsystematically and tentatively—to those elements of Poincaré's historical background which, though relevant to his philosophy, had been largely ignored. This applies, above all, to the philosophy of the physics of the principles (Lagrange, Poisson, Fourier, Cauchy in France) whose British representative, Hamilton, I also link—hypothetically (perhaps through Larmor and Whittaker)—with Ramsey (Essay 2). Similarly, though Poincaré wrote only *à propos* of Joseph Larmor's physical theories, he did pursue in detail Larmor's programme of analysing formal similarities between alternative theories of optics and electricity with important consequences for his epistemology (Essays 2 and 5). Gabriel Koenigs's kinematics, based on the theory of transformations, is in itself historically interesting (one of his definitions reads: "Absolute velocity is the geometrical sum of relative velocity and the velocity of entrainment", where "absolute" and "relative" have "a purely conventional sense", 1897, p. 82). Poincaré saw in "Koenigs's theorem", to which he refers repeatedly, the justification of the conventionalist thesis of the multiplicity of theoretical explanations for any set of observational data in mechanics (Essay 5). To my knowledge, this has never been discussed by critics.

Advocates of so-called incommensurability thesis in contemporary philosophy of science have failed to see its anticipation in the Poincaré–LeRoy dispute over the "universal invariant" or in Ajdukiewicz's radical conventionalism (Essay 4). This is presumably attributable to the circumstance that philosophers today do not expect to find much of relevance or in common with conventionalists whom they have come to see as total outsiders and whose writings they therefore rarely read. And yet, though conventionalist philosophy is in its method different from the contemporary socio-historical approach, many of its tenets and ideas are strikingly close to those which dominate current philosophy of science. This is due at least partly to the fact that both conventionalism and contemporary philosophy of science had their starting-point in the critique of traditional empiricism.

My own interest in the role of conventionalists in the history of ideas, originally due to personal contacts with Ajdukiewicz, was strongly enhanced when I noticed the complexity of their insider-outsider perspective (to use again Robert Merton's terminology) which defies epistemological stereotypes and sheds new light on the evolution of 20th century philosophy of science. Poincaré, for example, shared with the empiricists the belief that experiment is the final arbiter in the question of scientific truth *but* with the rationalists he shared the belief in the revealing power of mathematical intuition, which can often anticipate empirical results; as one of the originators of relativity theory he clearly saw the transient nature of the assumptions of classical mechanics, *but*, unlike Einstein and Minkowski, he did not believe that space-time is an absolute reality, uniquely imposed upon us; his "conventions", especially if understood as "terminological conventions" may be seen as anticipating the idea of "meaning-postulates", fundamental for logical empiricism, *yet* as a severe—and not always fair—critic of logicism and formalism he was clearly a precursor of their opponents. Ajdukiewicz's case is similar in many respects: he initiated in Poland in the early twenties the formal (syntactical) research into the foundations of logic and mathematics and stimulated the rise of logical semantics; in philosophy he was a representative (in his own words) of anti-irrationalism and saw in logic the method of philosophy *but* from the early twenties on he included in (meta-)logic what later came to be known as pragmatics; on this basis he conceived the programme of an understanding (humanistic) methodology of mathematics and science. Many of his results concerning language (syntactic connection, interrogative sentences, etc.) turned out to be directly applicable to ordinary language and have given rise, among other things, to now well-known Ajdukiewicz grammars. In the early thirties he surprised his Viennese and Berlin colleagues by a series of papers (in *Erkenntnis*) on radical conventionalism claiming, among other things, that revolutionary changes in science were not cumulative but disruptive and that, therefore, theories such as classical, relativistic and quantum mechanics were not comparable among themselves with respect to truth, content, confirmation, etc. Though in the dispute over the intertranslatability of sufficiently rich languages Ajdukiewicz took LeRoy's side—whence his *radical* conventionalism (Essay 4)—his interest was mainly focussed on Poincaré's philosophy. This came about in, roughly, the following way. Epistemological views similar to Poincaré's had been advocated in 1912 by Jan Łukasiewicz in his article "Creative elements in science" (1912, reprinted in 1915, 1934, 1961; English translation in 1970). The main point of that article was to criticise the traditional–empiricist doctrine, according to which science consists in the passive recording of facts, and to argue that throughout scientific knowledge there are creative elements contributed by the mind. Łukasiewicz's position was neither orthodox Kantian, in terms of fixed or evolving *a priori* categories, nor was it neo-Kantian in the sense of Lange's or Vaihinger's fictionalism. Though he compared science to art, like Poincaré he attributed a cognitive role to the creative elements in science. Łukasiewicz's article had a considerable influence on Polish philosophy of science in the first half of the 20th century, partly indirectly and partly through the writings of Ajdukiewicz and T. Kotarbinski. As regards Ajdukiewicz, a one-year visit to Göttingen University in 1913, five years after the "Poincaré–Festspiele" there (see Essay 5), prompted him to write an habilitation dissertation on the concepts of proof and existence in mathematics, in which he analysed critically Hilbert's and Poincaré's foundational views and distinguished the formal, semantical and pragmatic properties of axiomatic systems. Though through education, family background and language (he was bi-lingual) Ajdukiewicz had been much closer to the Austrian intellectual life, in his research after the Göttingen visit and after World War I, he followed the long-established Polish tradition of taking special interest in the philosophical developments in Western Europe, France and Britain in particular, in order to avoid regionalism and to counterbalance the influence of the German-speaking neighbours. This tradition explains, partly at least, why the philosophy of the French new critique of science came to be much better known in Poland than, for example, in Britain.

Jerzy Giedymin
University of Sussex

Acknowledgements

ACKNOWLEDGEMENTS and thanks are due to the following for the use of my previously published articles or parts thereof as stated separately in each essay: The Editor of *Studies in History and Philosophy of Science*, Vol. 8, No. 4, 1977; Cambridge University Press, *Prospects for Pragmatism*, ed. D. Hugh Mellor, 1980; D. Reidel Publishing Co., *Boston Studies in The Philosophy of Science*, XXXIX, 1976; The Editor of *The British Journal for The Philosophy of Science*, Vol. 24, 1973. Acknowledgement is also due to Peter Owen Ltd. for the use of a quotation from pp. 173–4 of *The Treasury of Mathematics*, Vol. 2, edited by Henrietta Midonick.

Contents

1. On the Origin and Significance of Poincaré's Conventionalism — 1
2. The Physics of the Principles and its Philosophy: Hamilton, Poincaré and Ramsey — 42
3. Duhem's Instrumentalism and its Critique: A Reappraisal — 90
4. Radical Conventionalism, its Background and Evolution: Poincaré, LeRoy and Ajdukiewicz — 109
5. Poincaré and the Discovery of Special Relativity — 149
Appendix: Logical Comparability and Conceptual Disparity between Newtonian and Relativistic Mechanics — 196
Bibliography — 206
Name Index — 217
Subject Index — 221

1

On the Origin and Significance of Poincaré's Conventionalism*

1. Programme for an Adequate Account of Poincaré's Philosophy of Science

READING Poincaré's philosophical writings is both exciting and frustrating. It is exciting for many reasons, not the least of which is the fact that he has anticipated so many of the problems relevant to contemporary philosophy of science though his solutions need not always be our solutions. It is frustrating because his writings abound in puzzling passages some of which do not seem to fit together. We know, of course, that Poincaré has not left a systematic exposition of his philosophy. He has produced a number of public lectures and articles written over a period of time, sometimes polemical in nature but more often than not addressed to authors unnamed. The majority of his philosophical articles appeared during his life-time in three collections, *La Science et l'Hypothèse* (1902), *La Valeur de la Science* (1905), *Science et Méthode* (1908). A fourth volume, entitled *Dernières Pensées* appeared posthumously in 1913; Poincaré died in 1912; he was born in 1854.[1] Apart from the fact that the mentioned collections do not include all of Poincaré's writings of a philosophical nature, it is almost impossible at the moment to appreciate the philosophical relevance of his voluminous works in pure and applied mathematics. It is not surprising, therefore, that Poincaré's philosophy of science has been subject to many disputes and that it has been interpreted in many mutually incompatible ways. Even the name of conventionalism given to it by some commentators is strongly disputed by others.

An adequate account of a doctrine in the philosophy of science is usually expected to answer at least the following questions. What is the origin and import of the doctrine? What problem(s) concerned with science or mathematics was the doctrine designed to solve? How does it differ from its main rivals? What effect has it had on further development of philosophy, science or mathematics?

However uncertain and controversial the rest may be, we know at least that Poincaré's philosophy originated from his research in the foundations of geometry. His first philosophical doctrine was published in 1887 when he was thirty-three years of age. It appeared in the form of brief epistemological comments appended to a mathematical memoir concerned with the foundations of geometry (Poincaré, 1887). Since in this first version of his philosophy of geometry Poincaré proclaimed the choice of a metric geometry to be similar to the choice of a coordinate system (e.g. Cartesian, polar, etc.) and in his later writings on the subject he referred to the axioms of geometry as "definitions in disguise" or conventions, quite justifiably the doctrine was given the name of *geometric conventionalism*, though Poincaré himself never used this appellation. It is worth noting at this point perhaps that conventionalist ideas in a broad sense appeared at the time (or a little earlier) independently of the research into the foundations of geometry in the writings of Friedrich Lange and Friedrich Nietzsche (cf., for example, his posthumously-published fragment "On truth and lie in an extra-moral sense"). At any rate, Poincaré's geometric conventionalism was the result of the conviction, widespread at the time among mathematicians and philosophers, that the discovery of non-Euclidean geometries, strengthened later by meta-geometrical results as well as by research into the origin and functioning of space perception in man and animals emanating from the theory of evolution, physiology and psychology of perception, made the orthodox Kantian view of space and geometry untenable.[2] Poincaré's geometric conventionalism was, therefore, designed to replace the Kantian view which had been undermined by these developments in mathematics and in biological science. What distinguished Poincaré from many of his contemporaries was the fact that having rejected the Kantian view of geometry and space, he did not seem to have embraced a straightforward empiricist philosophy. For in his 1887 article he denied that the axioms of geometry were "experimental facts" and in his later writings he vigorously criticised what he called "geometric empiricism" without, however, disclosing the names of any "geometric empiricists". The problem of the identity of "geometric empiricists" has become almost as intriguing to Poincaré scholars as the problem of the "Dark Lady" of Shakespeare's Sonnets to Shakespearian scholars. The very import of Poincaré's geometric conventionalism has become a subject of many interpretative disputes. The task of historians of ideas and of commentators has been made the more difficult by the fact that Poincaré retained the Kantian view of the foundations of analysis and refused to regard the axioms of arithmetic as an implicit definition of primitives, i.e. as conventional.[3]

In articles published in the last decades of the nineteenth century and at the beginning of the twentieth century Poincaré *extended the idea of conventionality and of convention* to his analysis of the *measurement of time*[4] and to what he called *the principles of physics*, e.g. the law of inertia, conservation principles, Hamilton's principle, the principle of relativity, etc.[5] This is the basis for extending the name of conventionalism to cover his whole epistemology and philosophy of science. I shall occasionally refer to the latter doctrine as *generalised conventionalism* to distinguish it from geometric conventionalism.

Disputes concerning the import of generalised conventionalism surrounded its birth and never relented. In a series of articles "Science et philosophie" (LeRoy, 1899), "Un positivisme nouveau" (LeRoy, 1901), and "La science positive et les philosophies de la liberté" (LeRoy, 1900–1), Edouard LeRoy, a pupil and ardent follower of both Henri Bergson and Henri Poincaré, combined elements of Bergsonian irrationalism, evolutionism and voluntarism with *extreme conventionalism* (*"nominalism"*) to argue for a "spiritualist philosophy of freedom" and for the superiority of religious experience over science. In "Un positivisme nouveau" he attributed his extreme form of conventionalism to the ideas developed by the representatives of so-called "nouvelle critique des sciences", among whom he mentioned E. Boutroux, Poincaré, Duhem, Milhaud, Wilbois, and himself. According to LeRoy, the basic tenets of the philosophy of the "new critique of science" were, firstly, pan-theoreticism, i.e. the view that there is no essential difference between so-called facts and theories, facts being theoretical constructs; secondly, the claim that all scientific laws and theories are conventions, hence so are all facts ("facts are created by scientists"); if conventions change, nothing remains invariant. Thirdly, science has meaning only from the point of view of life from which it emanates; hence it cannot be understood as a purely intellectual enterprise but only from the point of view of "action".

Not surprisingly, Poincaré felt it necessary to repudiate this "unintended consequence" of his conventionalist philosophy. In the preface to and in one of the chapters of *Science and Hypothesis* (1902), and in an article entitled "Sur la valeur objective des théories physiques",[6] he pointed out that LeRoy's nominalism as a philosophy of action was self-defeating; for if science is purely conventional, then it cannot serve as a basis of action, and if it *can* serve as a basis for action, then it cannot be purely conventional. Moreover, Poincaré affirmed against LeRoy that scientific facts were merely common-sense facts translated into a specialised, technical language of science, hence they are not freely created by scientists; finally, he claimed that under the changes of theoretical conventions there is an *"invariant"* with respect to which otherwise divergent theories are *comparable* or even *intertranslatable*.

The dispute between Poincaré and LeRoy provoked Pierre Duhem firstly to claim priority for the view of science which LeRoy attributed to all representatives of "the new critique of science" (though he dissociated himself from LeRoy's metaphysics), secondly, to express the belief that Poincaré's critical reply had been written with him (Duhem) in mind, though his name was not mentioned. Duhem also took this opportunity to question the tenability of Poincaré's view of scientific facts as common-sense facts translated into a scientific language; facts have no meaning independent of theories, he declared, and there is no one–one relation between facts and theories.[7]

Duhem's intervention in the dispute highlights the difficulties with which one is faced in interpreting texts: to understand Poincaré's writings one may have to disentangle first the mysteries of the philosophy of LeRoy, Duhem and many others of their contemporaries; but one cannot understand the latter without understanding Poincaré. We seem to be faced with the prospect of having to save the phenomena (given in texts) in terms of a combination of several hermeneutic circles.

In spite of the fact that Poincaré repudiated nominalism, i.e. extreme conventionalism, many later philosophers, e.g. Karl Popper[8] (1959: pp. 78–9; 1963: pp. 74, 99, 104) still blame him for holding the view that all scientific laws and theories are nothing but conventions, or formal instruments which cannot be refuted by experiments. At the same time, other commentators blame Poincaré for being a falsificationist and an inductivist and for regarding the principle of relativity as only a tenative, factual hypothesis (e.g. Holton, 1974).[9]

Strangely enough, in the disputes over the import of conventionalism no attempt was ever made to relate that doctrine to Poincaré's *main philosophical aim or concern* which, I think, was to examine changes in science and to show that there is continuous progress in spite of rapid apparently disruptive changes in theories. Poincaré expressed this concern both in Chapter X of *Science and Hypothesis* and in the preface to *The Value of Science*. In the latter he said:

> The advance of science is not comparable to the changes of a city, where old edifices are pitilessly torn down to give place to new ones, but to the continuous evolution of zoological types which develop ceaselessly and end by becoming unrecognizable to the common sight, but where an expert eye finds always traces of the prior work of the centuries past. . . .[10]

And in *Science and Hypothesis* he argued that though physical theories have been undergoing frequent transformations, yet certain features remain invariant: either some of the observational consequences or the type of formal structure of theories. What *does* change rapidly and radically are metaphors used by scientists such as various models of the atom, or the wave and the particle pictures of light, etc. These metaphors, though very important for the development of science do not express knowledge, for all that we can know are the relations between observable bodies or those relations between unobservables which have observable effects; reality is knowable only up to the observational equivalence and up to the isomorphism of the theoretical principles of rival theories.[11]

I will conclude this outline of some of the disputes over the import of conventionalism by proposing a list of requirements for a satisfactory account of Poincaré's philosophy of science:

(a) it should explain the origin and import of geometric conventionalism;

(b) it should clarify the relationship between geometric conventionalism and Poincaré's general epistemology and philosophy of science;

(c) it should decide whether the strong (nominalist) or the weak (moderate-empiricist) version of conventionalism (or neither) was the intended one;

(d) in solving the previously mentioned problems it should be guided by the hypothesis that in his writings Poincaré was pursuing the following dominant philosophical aim:

To show that progress in objective scientific knowledge is possible in spite of rapid, apparently disruptive changes in science, in spite of fashionlike changes in mathematical interests and viewpoints, in spite of the untenability of the two main antisceptical traditions, viz. orthodox rationalism and orthodox empiricism, in spite—finally—of the claims to the contrary voiced in many irrationalist, subjectivist and relativist doctrines which proliferated at the time.

I do not pretend to be able at present to give an account of conventionalism which satisfies all the listed requirements. In the rest of this essay I hope, however, to make a contribution to the programme, just outlined, of research into Poincaré's philosophy. I shall first report briefly what I think is the most interesting interpretation of geometrical conventionalism available, viz. the one given by Grünbaum, which traces geometric conventionalism back to Riemann's theory of manifolds and foundations of geometry. Then I shall comment on two of its main points in the light of Poincaré's writings and offer tentatively an hypothesis concerning the origin of geometric conventionalism which reflects on its import and links it with generalised conventionalism, i.e. Poincaré's philosophy of science.[12]

2. Geometric Conventionalism as an Epistemological Elaboration of the "Riemann–Poincaré Principle of the Conventionality of Congruence"

Adolf Grünbaum in his numerous contributions to the philosophy of space and geometry[13] traces geometric conventionalism back to Bernhard Riemann's theory of manifolds and foundations of geometry outlined in Riemann's celebrated *Habilitationsvortrag*.[14] "Poincaré's entire conventionalism in regard to metric geometry is a straightforward epistemological elaboration of Riemann's conception of the metric amorphousness of the spatial manifold" writes Grünbaum on one occasion,[15] and, on another, in a similar vein, "The central theme of Poincaré's so-called conventionalism is essentially an elaboration of alternative metrisability whose fundamental justification we owe to Riemann, and *not*... the *radical* conventionalism attributed to him by Reichenbach...."[16]. Judging from these quotations, Grünbaum's claim seems to be both of historical and logical nature. At any rate, the basis of geometric conventionalism on this view is what Grünbaum calls "the *Riemann–Poincaré principle of the conventionality of congruence*". It is the statement to the effect that since *physical space* as a continuous manifold is *metrically amorphous*, i.e. has *no intrinsic metric*, it—or any part of it—can be *metrised in many different ways* on the basis of various extrinsic standards. We may use, for example, a measuring rod as such a standard. The results of spatial measurements will be statements about the relations between the measuring rod and objects to be measured. Since the rod has to be transported, the question arises whether or not it is itself invariant under transport. This question turns out to be a non-empirical one. It can only be decided by convention and the latter will affect our definition of "congruence" of line-segments. If we adopt the convention that the rod's length is invariant under transport, i.e. that the rod is rigid, then this amounts to adopting *a coordinative definition* for the congruence of line-segments (a line-segment is congruent with our measuring rod if its end-points coincide with the end-points of the rod) and to choosing as our congruence class the set of all line-segments which are assigned equal length by the function:

$$ds = \sqrt{dx^2 + dy^2}$$

in accordance with the Pythagorean theorem. The geometry associated with this definition of congruence is Euclidean; the geodesics will be Euclidean straight lines. However, we can equally well metrise space or part of it by adopting a different convention concerning our measuring rod. We may decide to regard the rod as *dependent* in some specific way on its *position and orientation*. Accordingly, our coordinative definition will no longer be in terms of the coincidence of the line-segment with the end-points of the rod. As our distance function we may use, for example, the Poincaré function:

$$ds = \sqrt{\frac{dx^2 + dy^2}{y^2}}$$

The geometry associated with the Poincaré metric is non-Euclidean (hyperbolic); its geodesics will be a family of hyperbolically related lines whose Euclidean status is that of semicircular arcs. It follows that the same portion of space may be metrised in different ways. This *thesis of alternative metrisability* in terms of differential geometry corresponds to the *thesis of formal* (syntactical) *intertranslatability of alternative metric geometries* on the basis of a suitable "dictionary". The choice of a congruence class amounts to the choice of a metric geometry. So far, there is no question of truth or falsity involved. However, *once a convention* how to measure the length of line-segments *has been laid down*, the question which geometry is true of the physical space, becomes an *empirical question*, to be decided by actual spatial measurements, i.e. by experience.

The following seem to be the most important epistemological tenets of geometric conventionalism according to Grünbaum.

1. The principle of the conventionality of metric geometry which amounts to the Riemann–Poincaré principle of the conventionality of congruence is primarily a statement about an important structural property of physical space, viz. about its metric amorphousness and hence about its alternative metrisability.[17]

2. Only derivatively is the principle of the conventionality of congruence a statement about the intertranslatability of the languages of different metric geometries.

3. The results of spatial measurements are statements about relations between the standards of measurement and the objects measured; the choice of a distance function is essential for the very existence of metric relations not only for our ability to ascertain them.

4. The assumption that the standard is invariant under transport (or its negation) is not an empirical statement, it is conventional.

5. Once a convention concerning the method of measuring length has been made, the question whether physical space is Euclidean or not is an empirical one.

Grünbaum's interpretation of geometric conventionalism thus seems to give simple answers to at least two of my questions. The origin of geometric conventionalism is in the Riemannian theory of manifolds and its main import is the statement that physical space as a continuous manifold is metrically amorphous and can be metrised in different ways. The intended meaning of geometric conventionalism on this interpretation is evidently not the "strong" but the "weak" or moderate-empiricist version: pure geometry is seen either as an uninterpreted calculus or as an implicit definition of primitives; physical geometry is an empirical, testable theory. The only question that remains unanswered concerns the relationship between geometric conventionalism and the rest of Poincaré's epistemology and philosophy of science. Grünbaum himself, so far as I know, has not considered this question in his writings. However, if we were to follow his line of thought and if Poincaré's epistemology and philosophy of science were to be regarded as a generalisation of geometric conventionalism, then the former should be justifiable in terms of an ontological or physical statement which would correspond to the thesis of the amorphousness of physical space. It is not easy to find such a statement to justify generalised conventionalism unless it be the statement that *there is no natural classification*, understood either in the ordinary sense that there is no intrinsic way of classifying phenomena (objects, etc.) or else in the sense of the denial of Pierre Duhem's doctrine that there is a "natural classification", i.e. a unique, true description of Nature which science approximates in its development. Of course, one can also accept Grünbaum's interpretation of geometric conventionalism and claim that there is no generalised conventionalism, i.e. the relationship between geometric conventionalism and the rest of Poincaré's epistemology and philosophy of science is only loosely based on some vague concept of convention.

An Alternative View

If we confront Grünbaum's account of geometric conventionalism with Poincaré's writings, then—I think—several doubts arise. Firstly, if the so-called Riemann–Poincaré principle of the conventionality of congruence is primarily a statement about a structural property of physical space, viz. its continuity, hence metric amorphousness, then why does Poincaré seem to have avoided the use of the very concept of physical space? Why did he claim that experiments have no reference to space but only to bodies? Why did he say that space and time are not imposed on us by Nature, it is we who impose them on Nature?[18] How did he explain the conventionality of the number of dimensions of space? Why did he claim that the assumptions of the continuity or discreteness of matter are merely *indifferent* hypotheses from which nothing follows about the actual structure of matter? Secondly, if geometric conventionalism was an elaboration of Riemann's insights, and in epistemological terms amounted to moderate empiricism (later developed, for example, by Reichenbach and Carnap), according to which pure geometry is either uninterpreted or conventional and physical geometry is an empirically testable theory, then why did Poincaré criticise repeatedly and reject geometric empiricism? What did he mean by geometric empiricism? Who were geometric empiricists whom he criticised? Let us first consider some of his views concerning space.[19]

Poincaré's criticism of Kant's philosophy of space and geometry is based on the distinction between geometrical space and representative space (visual, tactile and motor). The two have different properties, for example, the former is continuous, homogeneous, infinite, etc., while the latter is not. We have invented geometrical space, Poincaré argues, to make consistent thinking in spatial terms possible; we do not represent to ourselves individual objects in geometrical space, rather we reason *as if* they were located in that space. Representative space is the product of genetics, of natural selection and of specific physical conditions in which we have evolved; all these factors are contingent. It follows from these claims that geometric space is not an *a priori* form of sensibility and representative space cannot be the basis of synthetic *a priori* truths. As we see, neither the idea of physical space nor the distinction between pure and physical geometry occur in Poincaré's criticism of the Kantian view.

In Chapter V ("Geometry and experiment") of *Science and Hypothesis* Poincaré affirmed that "Experiments teach us the relations of bodies to one another. They do not and cannot give us the relations of bodies and space, nor the mutual relations of different parts of space..." (p. 79) and "Experiments have reference not to space but to bodies" (p. 84). If we observed the movements of two sets of bodies and discovered that the structure of one is (approximately) that of the Euclidean group while the structure of the other is that of a non-Euclidean group, we must not conclude in the former case that space is Euclidean and in the latter that space is non-Euclidean (p. 83). No conclusions about space follow from either of the mentioned observations. Significantly, Poincaré excluded both in his 1887 article and in *Science and Hypothesis* (Part II, Chap. III, "Non-Euclidean geometries", p. 47) from his considerations of the foundations of geometry all Riemannian geometries in which the "axiom of free mobility" is not valid, i.e. all geometries of surfaces with variable curvature. It seems plausible to conjecture that he would also reject any conclusion based on measurements to the effect that physical space has variable curvature. For physical space in this sense would be a counterpart of Newton's absolute space which Poincaré rejected (except as a convention) and the view that it is knowable through experiments would amount to geometric empiricism, a doctrine to which he opposed his own.

Even if we agree with Grünbaum that the passages quoted here from "Geometry and experiment" (Poincaré, 1902, Chap. V) express merely Poincaré's *relational* view of space, viz. the view that the outcomes of spatial measurements concern relations between measuring rods or the velocity of light and objects measured (congruence relations, or the time it takes light to travel between two points), one would face the problem of finding an explication of Poincaré's conventionalist view of the dimensionality of space, in addition to Grünbaum's explication of the conventionality of metric geometry.

In "On the foundations of geometry" (1898a: pp. 21–4, 31–2) Poincaré argued—following Plücker—that the dimensionality of space depends on the choice of a space element, i.e. point (the transformations of a rotative subgroup—three dimensions), straight line (the transformations of a helicoidal subgroup—four dimensions), straight line and a point on it (the transformations of a rotative sheaf—five dimensions), etc. The choice is guided by the considerations of simplicity (1898a: p. 24) and is not imposed uniquely by experience; in this sense it is conventional.

In *Science and Method* (Chap. I "The relativity of space") Poincaré outlines his evolutionary epistemology with respect to space and geometry. It is an evolutionary epistemology both in the sense that human knowledge is seen in it as part and outcome of the process of biological evolution of organisms and in the sense that it appeals to the principles of the Darwinian theory of evolution. Poincaré gives there an account of how animals, including men, construct "local and extended space" and how man constructs on this basis "great space". Then the following claims are made: "the characteristic property of space, that of having three dimensions, is only a property residing, so to speak, in the human intelligence...." (p. 112). There are animals that see the world in two dimensions and there might be beings who would think (for practical purposes) in hyperspaces, though they might not have as good a chance of survival as we have. It is certainly possible, so Poincaré continues, to translate physics into the language of hyperspaces, as Hertz has done in his mechanics (and Gibbs in statistical mechanics in which the "phase space" is of 6N dimensions). The fact that we have evolved a three-dimensional view of the world and have survived so far is no proof that space has three dimensions any more than the theoretical success of Hertz (or Gibbs).

I conclude from these considerations that Poincaré used the concept of space as a physically uninterpreted one. It is, therefore, uncertain if he would see the justification of geometric conventionalism in a statement about the structural properties of physical space.[20] If one were to find a common denominator for the conventionality of the metric, of dimensionality of space and of physical theories, it would have to be an epistemological thesis, for example the claim of the existence of observationally equivalent theoretical systems. Now it is understandable that the basis of the conventional nature of the dimensionality of space may be different from that of the metric (congruence), but Poincaré included the problem of the dimensionality of space in his critique of geometric empiricism (1898a: p. 24; 1902, 1952: pp. 68–70, 87–8). He wrote, for example: "We might be tempted to conclude that experiment has taught us the number of dimensions of space; but in reality our experiments have referred not to space, but to our body and its relations with neighbouring objects. What is more, our experiments are exceedingly crude..." (1902, 1952: p. 87). Geometric conventionalism, as Poincaré's replacement for geometric empiricism, should—therefore—comprise and explain both the conventionality of the metric and of the number of dimensions. One may wonder what is their common source, so to speak. One wonders also whether the continuous nature of the number manifold taken as the mathematical model of space (a model "which we carry about in us"), is necessary or sufficient for the conventionality of the metric and whether Poincaré would see in it a structural property of physical space. Doubts concerning the latter may arise from an apparently analogous argument in Chapter IX "Hypotheses in Physics" of (1902) in which Poincaré distinguished "different kinds of hypotheses" depending on how ready or reluctant we should be to part with them.

"In most questions the analyst assumes, at the beginning of his calculations, either that matter is continuous, or the reverse, that it is formed of atoms. In either case, his results would have been the same. On the atomic supposition he has a little more difficulty in obtaining them.... If, then, experiment confirms his conclusions, will he suppose that he has proved, for example, the real existence of atoms?" (1902, 1952: p. 152).

Accordingly, Poincaré classified both the assumption of the continuity and of discontinuity of matter, made as part of a mathematical model, among the hypotheses which he gave the name of "indifferent hypotheses" and which he regarded as non-descriptive, interchangeable conventions. On the other hand, in the (1898a) article on the foundations of geometry, Poincaré argued that the assumption of discontinuity (i.e. "each displacement forms part of a sheaf formed of all the multiples of a certain small displacement far too small to be appreciated by us...", p. 15) or of "semi-continuity" ("... the different displacements of the sheaf would be, so to speak, commensurable with one another...") in geometry would be incompatible with other properties "which experience has' led us to adopt for the fundamental group of geometry...", viz., "... The group contains an infinity of subgroups, all *gleichberechtigt*, which I call rotative subgroups. The rotative sub-groups have a sheaf in common which I call rotative and which is common not only to two but also to an infinity of rotative subgroups. Finally, every small displacement of the group may be regarded as the resultant of six displacements belonging to six given rotative sheaves. A group satisfying these conditions can be neither discontinuous nor semicontinuous..." (1898a: pp. 37–8). As we know, however, according to Poincaré this does not mean that a group with such properties is *imposed* on us by experience (if it did, conventionalism would be indistinguishable from geometric empiricism). All our experiences with displacements are equally compatible with the assumption that the group (geometrical space) is neither continuous, discontinuous nor semi-continuous but has some different, more complex structure. The reason why "we reject it, or rather we never think of it" is because this would be a more complicated hypothesis (1898a: p. 38). Observational equivalence and structural similarity is here the limit of our knowledge of the structure of space, as it is in any other case of theoretical knowledge. This appears to me to be the most fundamental tenet of conventionalism and perhaps the common source of the conventionality of the metric, of the number of dimensions of space and of all experimentally indistinguishable theoretical descriptions. Empiricism, geometrical or otherwise, which Poincaré criticised and rejected, apparently ignored the existence—for any given set of data—of many empirically adequate hypotheses and—more generally—of observationally equivalent ones. This brings us to *the other controversial point* in Grünbaum's interpretation of geometric conventionalism, viz. the claim that *Poincaré's view of the status of geometry* was essentially *empiricist*, strictly speaking moderate-empiricist, similar to the view later developed, for example, by Reichenbach and Carnap. This implies the distinction between pure and applied (physical) geometry. The former is either an uninterpreted calculus or else an implicit definition of primitive terms, i.e. it is a set of "meaning-postulates" (to use Carnap's term), and as such analytic. The latter, on the other hand, is a physical theory obtained from a calculus by laying down coordinating definitions for some of the primitive terms and as such it is experimentally testable. In particular, when a coordinating definition for congruence of line-segments has been adopted, the question whether the resulting system of geometry (Euclidean or non-Euclidean) is true of the physical space, becomes an empirical question to be decided by actual spatial measurements. However, if this moderate-empiricist view of geometry is attributed to Poincaré, then one has to explain why it was necessary and how it was possible for him to criticise and reject "geometric empiricism" from such an empiricist standpoint. This in turn leads to the question of the identity of geometric empiricism, and of geometric empiricists.

The best known of Poincaré's *arguments against "geometric empiricism"*—and—strictly speaking—the only one discussed by Grünbaum—is Poincaré's *apparent criticism of the use of parallactic measurements in a crucial experiment to decide between alternative metric geometries*. For brevity's sake I shall refer to Poincaré's criticism of the parallax experiment as *"the parallax argument"*. The relevant text is in *Science and Hypothesis* (Chap. V "Experiment and geometry", pp.

72–3):

> If Lobatchevsky's geometry is true, the parallax of a very distant star will be finite. If Riemann's is true, it will be negative. These are the results which seem within the reach of experiment, and it is hoped that astronomical observations may enable us to decide between the two geometries. But what we call a straight line in astronomy is simply the path or a ray of light. If, therefore, we were to discover negative parallaxes, or to prove that all parallaxes are higher than a certain limit, we should have a choice between two conclusions: we could give up Euclidean geometry, or modify the laws of optics, and suppose that light is not rigorously propagated in a straight line....

Many commentators (including Reichenbach, Weyl, Nagel) have taken this passage to mean that, according to Poincaré, Euclidean geometry as physical geometry can always be retained in the face of an apparently negative experimental outcome, if we are prepared to make suitable adjustments in the associated physics (optics). Like the so-called *Duhemian thesis*, the parallax argument seems to affirm the ambiguity of experimental testing of physical geometries, i.e. the impossibility of a crucial experiment which would unequivocally decide between alternative metric geometries, hence the possibility of always rescuing the preferred geometry from experimental falsification. Popper, who classified both Poincaré and Duhem as extreme conventionalists (nominalists), has introduced the term "conventionalist strategems" to refer to such methods of saving one's preferred hypothesis. By contrast, M. Hesse argues that Duhem's thesis is weaker than Poincaré's position under discussion, for the former denies the possibility of testing isolated hypotheses while the latter also denies the possibility of testing "total theoretical systems".[21]

Grünbaum rejects all these interpretations of the parallax argument and claims that the relevant passage in *Science and Hypothesis* merely affirms *alternative metrisability* of space, i.e. it reaffirms a direct consequence of the Riemann–Poincaré principle of the conventionality of congruence. Since the principle itself is perfectly compatible with the moderate-empiricist view of geometry, so is the parallax argument. Having chosen a method of measuring length of line-segments and the associated geodesics (e.g. light-rays) we can change our decision and re-metrise the relevant portion of space using a different method of measurement *without thereby changing the empirical content* of our system of physical geometry. All that is involved is the re-naming of some physical (optical) objects. There is, therefore, no question of the parallax argument being a special case of the Duhem thesis which allows the retention of a hypothesis by means of a change in the empirical content of the system under testing. In the Duhemian case the interdependence of geometry and physics is "inductive" (epistemological) while in Poincaré's case it is merely linguistic.

In support of his interpretation of the parallax argument Grünbaum makes use of *text-analysis* and also appeals to the general *historical context*. Firstly he points out a fact, overlooked by other commentators, that the quoted passage in *Science and Hypothesis* had been lifted verbatim from an earlier article entitled "Des fondements de la géométrie: À propos d'un livre de M. Russell" (1899)[22] in which Poincaré criticised some of Russell's views on geometry expressed in Russell's book *An Essay on the Foundations of Geometry* (1897). From that original context it is clear, so Grünbaum claims, that the parallax argument was used by Poincaré against the doctrine (held by Newton and also by Russell) of the existence of an intrinsic metric of physical space; in other words, the parallax argument was used to affirm alternative metrisability due to the absence of any intrinsic metric of space. To strengthen further the claim that Poincaré's criticism of so-called geometric empiricism was nothing but a criticism of the doctrine of the existence of an intrinsic metric, Grünbaum makes the following diagnosis of the historical context in which the parallax argument was used against geometric empiricism:

> ...at the turn of the century the Riemannian kind of empiricist conception of physical geometry...which we now associate with writers like Carnap and Reichenbach and which takes full cognizance of the stipulational status of congruence, had hardly secured a sufficient philosophical following to provide stimulus and furnish a target for Poincaré's polemic....[23]

Now in my own view the parallax argument must not be taken to imply that Poincaré denied the possibility of any empirical testing, i.e. it must not be understood as committing Poincaré to a nominalist, in this sense anti-empiricist position, for such an interpretation would be obviously in collision with Poincaré's explicit disavowal of nominalism in his polemic with LeRoy. I also agree with Grünbaum that the parallax argument implies the claim of alternative metrisability of space. However, I think that Grünbaum's arguments in support of this contention are questionable and that *the parallax argument implies more than alternative metrisability*. Since, in my view, both these points are essential to the problem of the origin and import of geometric conventionalism, I will discuss them in some detail.

Grünbaum's arguments in support of his interpretation of the parallax argument seem to me unconvincing for the following reasons. First of all, though he is right to refer the parallax argument to Poincaré's article (1889) against Russell, he does not mention the fact that in that article the argument appears in a passage which ends with these words: "Nothing authorises me to assume that Mr Russell's doctrine implies such a crude interpretation. I believe that he wanted to affirm other things, more subtle and special..." (p. 256). Secondly—and more importantly—Grünbaum does not mention the fact that suggestions to use astronomic—in particular parallactic—measurements to decide between alternative metric geometries had been made, for example, by N. Lobatchevsky in his *Untersuchungen zur Theorie der Parallel-Linien* (1840)[24] before B. Riemann's *Probevorlesung* (1854).[25] Both were, of course, known to Poincaré. The circumstance that Riemann's moderate geometric empiricism had been discussed as such by A. Cayley in his Presidential Address to the British Association (1884: pp. 6–9) could have provided a sufficient stimulus for Poincaré to take it for a target. Finally, Grünbaum disregards the fact that *Poincaré had criticised geometric empiricism twelve years before his polemic with Russell*, viz. in his 1887 article on the foundations of geometry which was clearly written with Riemann's memoir in mind[26] and which appeared four years after Cayley's Address.

The very title of Poincaré's 1887 article "Sur les hypothèses fondamentales de la géométrie" is a literal translation into French of the title of Riemann's memoir *Über die Hypothesen welche der Geometrie zugrunde liegen* (1854). The whole article is mathematical in nature (which presumably explains why it has been so rarely used in discussions of Poincaré's philosophy) except the last page, concerned with epistemological questions. Poincaré's *mathematical objective* was similar to Riemann's though more modest: to give a list of assumptions necessary and sufficient to develop metric geometries, except that (unlike Riemann) he confined himself to plane geometry. To leave no doubt as to the context in which his article should be seen, Poincaré begins the epistemological commentary with the following sentence: "Readers who have been willing to follow me up to this point, could not have failed to see that what went before was related to Riemann's celebrated memoir ... nor could they have failed to notice certain differences between our respective methods and results ..." (p. 214). The main mathematical differences between the two memoirs are, firstly, that Poincaré used the *group-theoretic approach* to geometry and, secondly, that he excluded all geometries in which the axiom of free mobility is not valid, i.e. he excluded all geometries of surfaces with variable curvature. *The epistemological problem* is introduced by Poincaré in the following way: "One could now pose the question what are these hypotheses? Are they experimental facts, analytic judgements or synthetic judgements *a priori*?" Now, Riemann in his memoir refers to the assumptions of geometry as either hypotheses or "simple facts" (*einfache Tatsachen*). Poincaré's "experimental facts" seems to be a translation of Riemann's "*einfache Tatsachen*".[27] His answer to the epistemological question is that the assumptions of geometry are neither synthetic *a priori* truths, nor analytic truths, nor experimental facts. Geometry is nothing but a study of a group. In this sense the truth of Euclidean geometry is not incompatible with the truth of Lobatchevsky's geometry, since the existence of one group is not incompatible with the existence of another group. From among possible groups we choose one to refer to it physical phenomena, just as we choose three coordinate axes to refer to them a geometric figure. The choice is determined by the simplicity of the group and by empirical considerations such as our knowledge of the existence of solids whose movements have roughly the same structure as that of the group chosen. Though geometric hypotheses are not experimental facts, our choice of those hypotheses is guided by the observations of certain physical phenomena.

The selected group is merely the most convenient one, hence to say that Euclidean geometry is true while Lobatchevsky's is false would make no more sense than to say that Cartesian coordinates are true and polar coordinates false.

I have reported the epistemological commentary contained in Poincaré's 1887 article for three reasons.

1. So far as I know, it was the first formulation of Poincaré's geometric conventionalism and as such important for any investigation of the origin of that doctrine.

2. It seems to reject not only the Kantian view of geometry but also the Riemannian view which Poincaré apparently regarded as empiricist, i.e. as implying that it is possible to decide on experimental grounds which geometry is true of physical space.

3. It suggests that the source of Poincaré's geometric conventionalism must be sought *not* in Riemann's insights and discussion of the foundations of geometry but elsewhere.

If my conjectures are correct, then—contrary to Grünbaum's view—*Riemann* (along with Gauss and Lobatchevsky) would be among Poincaré's geometric empiricists and as such the target of Poincaré's criticism from the conventionalist point of view. In fact, if one reads Riemann's memoir, one is struck by the conspicuous *absence* from his text of any clear traces of conventionalist ideas and terminology (words such as "choice", "decision", "arbitrary" let alone "convention", never occur; "simplicity" does occur but only in the way in which one could talk about relative simplicity of alternative empirical hypotheses); the tendency is obviously empiricist. The conventionality (in some sense) of congruence may have been implied by his claim of the amorphousness of physical space as a continuous manifold. However, even this is uncertain in view of the explicit suggestion made by Riemann that "the basis of metric relations must be sought... in the colligating forces that operate upon it (space)".[28] It would be hardly surprising if Poincaré had understood Riemann as *not* "taking full cognizance of the conventionality of congruence". If this was the case, then the target of Poincaré's criticism would be both Riemann's apparent failure to recognise the conventional nature of extrinsic criteria for metric relations as well as the idea that experience may discover the

Euclidean or non-Euclidean nature of physical space. Geometric empiricism (criticised and rejected by Poincaré) was—I suggest—the view that there is an intrinsic metric in space to be discovered by experience, or that there is no intrinsic metric in space itself but it can be discovered experimentally in "the colligating forces" which operate on space or, in general, the view that one can discover by experience alone whether physical space is Euclidean, non-Euclidean or with variable non-Euclideanism.

If Poincaré's conventionalism was not directly inspired by, and was *not* a simple epistemological elaboration of, Riemann's insights concerning continuous manifolds, then *what was its source and its significance?*

The main clue to this problem is, I think, in Poincaré's 1887 article, which was *inspired by Sophus Lie's theory of groups of transformations.* Poincaré's investigations into the foundations of plane geometry presented in that article relied on one of Lie's theorems to the effect that "If the position of a plane figure on a plane depends on a finite number of conditions, the number of those conditions cannot exceed eight". Poincaré quotes the theorem both in his 1887 article (p. 207) and in *Science and Hypothesis* (Part II, Chap. III "Non-Euclidean geometries", p. 47) to explain why in contrast to Riemann he considered only a relatively small number of non-Euclidean geometries, viz those which are compatible with the movements of an invariable figure, an assumption on which Lie's theorem is based.

Although it is, of course, well known that Poincaré was inspired in his research into the foundations of geometry by Lie's group-theoretic approach, Lie's writings have not—so far as I am aware—been examined as a possible source of inspiration of Poincaré's geometric conventionalism.

An examination of Lie's memoir "On a class of geometric transformations" (1871)[29] proves extremely illuminating not only from the purely mathematical but also from the epistemological point of view. In contradistinction to Riemann's *Habilitationsvortrag*, Lie's memoir contains explicit ideas which could clearly have suggested to Poincaré a *conventionalist philosophy of geometry (not only metric) whose emphasis is on the intertranslatability (in a sense) of various geometries.*

The very first sentence of Lie's memoir refers to *the philosophical view of geometry* which, I suggest, was *the source of the first* (1887) *formulation of Poincaré's geometric conventionalism*: "The rapid development of geometry in the present century has been closely related to and dependent on the philosophic views of the nature of Cartesian geometry, views which have been set forth in their most general form by Plücker in his earlier works..." (pp. 485–6).[30] Then Lie continues as follows:

> Cartesian analytic geometry translates any geometric theorem into an algebraic one and effects that the geometry of the plane becomes a representation of the algebra of two variables and likewise that the geometry of space becomes an interpretation of the algebra of three variable quantities.... Plücker has called our attention to the fact that Cartesian analytic geometry is encumbered by a two-fold arbitrariness. Descartes represents a system of values for the variables x and y by a point in the plane; as ordinarily expressed he has *chosen the point as element for the geometry of the plane*, whereas one could with equal validity employ for this purpose the right line or any curve whatsoever depending on two parameters.... Furthermore Descartes represents a system of quantities (x, y) by that point in the plane whose distances from the given axes are equal to x and y; *from an infinite number of possible coordinate systems he has chosen a particular one*. The progress made by geometry in the 19th century has been made possible largely because this two-fold arbitrariness in the Cartesian analytic geometry has been clearly recognized as such....

After this historical diagnosis, in which conventionalist terminology abounds ("two-fold arbitrariness", "has chosen... could have with equal validity have chosen", "translate") Lie explains that the new theories developed in his memoir "are based on the fact that any space curve involving three parameters may be selected as element for the geometry of space" (pp. 489–90) and that, in general, the choice of a geometry is, so to speak, an opportunist affair: one develops and uses a geometry which is *advantageous* or *convenient* for solving the problems at hand. Moreover, one takes advantage of the fact that *one geometry may be transformed into another*, for example, Plücker line geometry may be transformed by a contact transformation into a space geometry whose element is the sphere (this is one of the main results of Lie's article; F. Klein has proved that Plücker line geometry, hence also sphere geometry, illustrates the metric geometry of four variables). This means that the *problems* and *theorems* of such two geometries are, so to speak, *intertransformable*: a problem concerning spheres may be transformed into a problem concerning lines and solved, perhaps, in an easier way.[31] Let us call this insight concerning geometries *the Plücker–Lie principle of transformation*. It is, I think, one of the grounds of Poincaré's geometric conventionalism.

Lie's theory of geometric transformations is based on the *Poncelet–Gergonne theory of reciprocity*. If (x, y) and (X, Y) are Cartesian point coordinates for two planes, then the equation:

$$X(a_1x + b_1y + c_1) = Y(a_2x + b_2y + c_2) + (a_3x + b_3y + c_3) = 0$$

establishes a reciprocity between the two planes in the sense that, mutually, to the points of one plane correspond the right lines of the other; to the points of a given right line λ correspond the right lines that pass through λ's image point.[32]

Now it is significant that Gergonne was one of the first to notice (in 1826) and utilise in his exposition of projective geometry the *duality of geometric axioms* owing to which through *interchange of certain terms*, e.g. "point" and "plane", one can obtain valid (true) statements from other valid statements; for example, from "Two points determine a straight line" one obtains "Two planes determine a straight line". Duality may characterise terms and formulae *within one* axiomatic system or it may hold *between two* such systems. Duality is related to the fact that an axiom system merely establishes *relations between primitive terms* and, in this way, merely imposes restrictions on admissible interpretations of those terms. Interchange of terms may thus leave both the structure of the system (consequence relation) and its realisations unaltered. The method of exposition used by Gergonne in discussing duality in projective geometry is similar to the dictionary used by Poincaré, for example, in *Science and Hypothesis* (pp. 41–2) to translate Lobatchevsky's geometry into Euclid's.

It was also Gergonne who in the *Essai sur la théorie des définitions* (1818) introduced the idea of an *implicit definition* in contrast to an *explicit* (equivalence) *definition*. The idea was based on the *analogy with equations*. A sentence with one term whose meaning is unknown (or a new term to be introduced into the language) is analogous to an equation with one unknown; a set of axioms with n primitive terms is analogous to a set of n equations with n unknowns. The roots which satisfy the equations are analogous to the interpretations of primitives under which the axioms are true. *This is the origin of Poincaré's views of geometrical axioms as "disguised definitions"* (*Science and Hypothesis*, p. 50) and—in general—of the conventionalist idea of "terminological conventions", "meaning postulates", "axiomatic meaning rules", etc.[33]

Naturally, just as not every set of equations determines a set of values (roots) by which it is satisfied, not every axiom system may serve as an implicit definition of primitives. There are two *conditions* which every mathematical definition must satisfy: one has to prove the *existence* and *uniqueness* of the object to be defined. When applied to an axiom system as an implicit definition of its primitives, these conditions amount to the requirement for the proofs of consistency and categoricity. Both are important in order to understand why Poincaré regarded the axioms of all metric geometries as implicit definitions of their primitives and—in this sense—as terminological conventions but denied this status to the Dedekind–Peano axioms of arithmetic. In his 1887 article he claimed that while some of the assumptions of analysis are synthetic *a priori* truths, all geometrical axioms are conventions. It is only, however, in his *Science and Method* that he explicitly discussed his reasons for the distinction.[34] The consistency of geometries, Poincaré argued, may be proved relative to arithmetic (of real numbers; they can also be shown to be categorical). On the other hand, to prove the consistency of the arithmetic of natural numbers, e.g. of the Dedekind–Peano axiom system, one has to use the axiom of induction. Thus such a proof would involve a *petitio principii* and hence is impossible. We are, therefore, obliged to accept the principle of mathematical induction—without which mathematics is impossible—as a synthetic *a priori* truth. This means, however, that—in contrast to the axioms of geometries—the axioms of arithmetic must not be viewed as an implicit definition of arithmetical primitives, i.e. as conventional or as merely a more or less convenient language.[35]

From what has been said so far I draw the following hypothetic *conclusions* concerning the *origin* and *significance* of Poincaré's *geometric conventionalism*:

The origin or source of inspiration of Poincaré's geometric conventionalism is to be sought primarily in the research in geometry and in the associated philosophy of *Gergonne, Plücker and Lie* rather than of Riemann whom Poincaré regarded as one of the "geometric empiricists", along with Helmholtz and Lobatchevsky.

As regards the *significance or import of geometric conventionalism* of Poincaré, I suggest that it should be seen as embedded in the following view of geometry.

1. Geometries are complex *language-systems*; seen as axiomatic theories they are *implicit definitions* of their primitive terms, which means that the geometric primitive terms are interpreted only by postulates (axioms);[36]

2. Some of these languages are *intertransformable*, in this sense *intertranslatable*; this is taken advantage of to facilitate the solution of geometrical problems: by a suitable choice of geometry one may simplify the solution of a problem just as one may do so by a suitable choice of coordinates.[37]

3. "Space" has no physical interpretation; as a mathematical continuum it is amorphous and can be metrised in various ways on the basis of various conventions concerning "distance" or "congruence"; congruence of two figures means that they can be transformed one into another by a point transformation in space; congruence is an equality relation in virtue of the fact that displacements of figures are given by a group of transformations; there are three types of continuous groups in space which in a bounded region have the property of displacement: these groups correspond to the three metric geometries of Euclid, Bolyai–Lobatchevsky and Riemann.[38]

4. The role of experience in geometry is two-fold: geometrical concepts and assumptions *originate from experience*; from the status of idealised empirical generalisations geometrical assumptions are then elevated to the status of *conventional principles* or *terminological conventions*; moreover, in the *applications of metric geometry* we are *guided in our choice* of a system of metric geometry by its *simplicity* (in psychological, pragmatic and mathematical senses) and *convenience*

but also by *empirical considerations* relevant to simplicity and convenience, for example, by our knowledge of the existence in nature of solid bodies whose movements closely approximate the structure of the Euclidean group. This does not mean, however, that we test experimentally applied metric geometries; what would be the point of such tests, if "space" has no physical meaning? We do not even test metric geometries with respect to groups of physical bodies and their movements—rather we adjust our geometry to empirical findings by a suitable choice of the definition of "congruence" and by similar linguistic devices; in other words, with respect to metric geometries we adopt the *"nominalist attitude"* as a result of which they function as languages rather than as empirical theories, even in physical applications.[39] They remain "exact sciences", not subject to revision which would not be the case if they were tested.

The last point brings us back to our discussion of *Poincaré's criticism of the parallax experiment as a crucial experiment to decide between alternative metric geometries* and to the question how this criticism was related to Poincaré's rejection of "geometrical empiricism". I claimed at the beginning of the present chapter that though the parallax argument (i.e. Poincaré's criticism of the parallax experiment) *does* imply the claim of re-metrisability, as argued by Grünbaum, nevertheless, it implies more than that. It is time now to explain what this "more" is. I propose for this purpose the following *interpretation of the parallax argument* based on my account of Poincaré's philosophy of geometry.

(a) Poincaré subscribed to a *generalised version* of what is now often called the *Duhem thesis*;[40] let us call this generalised version the *Poincaré thesis*; it affirms that empirical testing is not possible without an experimental set-up which consists of two basic elements: a language and factual (empirical) hypotheses. The falsification of an empirical hypothesis may be avoided *either* by blaming the negative outcome of the experiment on one of the auxiliary hypotheses *or* by changing the language; the retention of Euclidean geometry by re-metrisation is an operation of the latter type.

(b) The Poincaré thesis implies that our *language is not fixed forever* but is, on the contrary, changeable and may even be changed in response to experimental findings: such a change of a language element does not amount, however, to *falsification* which *presupposes a fixed language*; it need not amount to any change in empirical content provided that the transition is to another language intertranslatable or observationally equivalent with the previous one, as is the case—according to Poincaré—with metric geometries.[41]

(c) Since falsification presupposes a fixed language and this means a limitation of our freedom to change our language in mid-stream, so to speak, a *decisional element* is involved; in this sense experience alone cannot impose any theory on us, the choice of a language and the decision to use a fixed language are also necessary.[42]

(d) If science as a whole is to be empirical and not purely conventional (nominalist), then each of the two sets of elements of the experimental set-up must not be empty; moreover, falsification must, at least sometimes, be possible which means that the language of science must at least sometimes be used as fixed.[43] On the other hand,

(e) It is convenient to use geometries as languages rather than as empirical theories; as such they are not tested and do not have to be kept fixed to make falsification possible.[44]

We can now ask the question whether—in the light of the proposed interpretation of geometric conventionalism—*Poincaré's philosophy of physics can be seen as "generalised conventionalism"*? My answer to this question is "yes", with the qualifications and in the sense to be specified presently.

One of the fundamental theses of generalised conventionalism is the *epistemological thesis of "the physics of the principles"*.[45] The principles of physics in question, like geometrical assumptions, are idealised empirical generalisations some of which have been *elevated to the status of conventions*.[46] The principles of mathematical physics (for example, the principle of conservation of energy, Hamilton's principle in geometrical optics and in dynamics, etc.) systematise experimental results usually achieved on the basis of two (or more) rival theories, such as the emission and the undulation theory of light, or Fresnel's and Neumann's wave theories, or Fresnel's optics and Maxwell's electromagnetic theory, etc. They express the common empirical content as well as (at least part of) the mathematical structure of such rival theories and, therefore, can (but need not) be given alternative theoretical interpretations. Principles usually survive the demise of theories and are responsible for the continuity of scientific growth.

The thesis of generalised conventionalism implies two further claims: *firstly*, that there has been a *growing tendency* in modern physics to *formulate and solve physical problems* within *powerful, and more abstract, mathematical systems of assumptions*, i.e. a tendency towards a more "rationalist" approach by comparison to gradual empirical approximations; in other words, there is more abstract mathematical thinking and less direct testing going on in physics than is dreamt of in empiricist philosophy; *secondly*, the role of conventional principles has been growing and *our ability to discriminate experimentally between alternative abstract systems* which, with a great approximation, save the phenomena *has been diminishing* (by comparison to the testing of simple conjunctions of empirical generalisations).

The principle of duality, mentioned before in connection with geometric conventionalism, was seen by many nineteenth century mathematicians and mathematical physicists as a universal law of nature which gives rise to dual systems not only of geometry but also of mechanics, optics, etc. The duality between Fermat's principle of least time and Maupertuis's principle of least action, both of which may be obtained from the law of optical path, was utilised by William R. Hamilton (an ardent Kantian, by the way) in his application of algebra to geometrical optics in his memoir, *A Theory of Systems of Rays*, and commented on in an abstract of that memoir in 1827. His wave mechanics was inspired by the associated parallelism between waves and particles. It is such dualities or "parallelisms" that Poincaré had in mind when in *Science and Hypothesis* (Chap. X) he pointed out that often though there is conceptual change (as, for instance, in the transition from Fresnel's optics to Maxwell's theory) the equations which express fundamental theoretical relations remain the same, for they are consequences"... of more general principles such as that of the conservation of energy and that of least action..." (p. 162) and reflect "a profound reality".

In other words, we have come to realise—as a consequence of this tendency towards mathematical abstraction—that *physical reality is* knowable only up to the observational (numerical-predictive) equivalence of alternative theoretical systems and up to the isomorphism of their theoretical postulates.[47] Theoretical systems in physics which are observationally and structurally indistinguishable are *mutually intertranslatable languages* (possibly in the holistic sense of the word). This important epistemological insight Poincaré owed partly to his own work as a mathematician and as a theoretical physicist, and partly to Kant's philosophy.

Generalised conventionalism—like geometrical conventionalism—was inspired by some of Sophus Lie's contributions to the theory of groups of transformations. In his applications of that theory to the study of differential equations, Lie has proved that all differential equations admit some groups, i.e. for every differential equation there is a group of (continuous) transformations which leave the form of the equation invariant; he has also shown that only when the group of the equation is found is it possible to integrate the equation by quadratures: thus the problem of finding the group or of integrating the equations by quadratures is one and the same. Any additional group of transformations admitted by a differential equation which one discovers simplifies the problem of integrating the equation. This illustrates the fact that what we have called *the Plücker-Lie principle of transformation* applies not only to geometries but to other mathematical theories, some of which have applications in theoretical physics. Of course, from the Lie-Klein viewpoint—adopted by Poincaré in his 1887 article—*geometries were studies of invariants under groups of transformations.* The "geometrisation" of physics which resulted from this approach made in turn *physics a study of invariants under groups of transformations* (the Galilean group, the Lorentz group, etc.). The principle of relativity, according to which the laws of physical phenomena are the same for a fixed observer as for an observer who has a uniform motion of translation relative to him is equivalent to the principle that all laws of physics are invariant under Lorentz transformation.

Some of the principles in physics are used to simplify the solution of problems in a straightforward sense. *Conservation principles* are so-called *first integrals* which facilitate the solution of dynamical problems; the problem of two bodies, for example, may be solved by the method of first integrals. Poincaré introduced also so-called integral invariants (dynamical magnitudes which remain constant during motion) which are analogous to first integrals in this respect.

How does conventionalism in this sense relate to Poincaré's view on progress in science? Although it may seem paradoxical, by emphasising the role of conventional elements in science, Poincaré sought to define the view that there is *continuous progress in objective scientific knowledge,* in spite of frequent, apparently disruptive changes.

The "paradox" disappears, however, if one realises that by "objective" Poincaré meant "intersubjective" ("what is common to many thinking beings, and could be common to all", *The Value of Science,* Introduction, p. 14) and by "conventional" he did *not* mean "arbitrary". Most importantly, among conventions he distinguished so-called *indifferent hypotheses* which can be replaced without the empirical content of theories being thereby affected. The developments in mathematics and in science as well as "the new critique of science" had shown that the orthodox rationalist (Kantian) and the traditional empiricist views of objectivity, rationality and certainty of scientific knowledge were no longer tenable. The basic principles of mathematics and of science turned out not to be synthetic *a priori* truths; facts turned out to be not as simple and objective as traditional empiricists claimed they were. One had to admit that both mathematics and science were man-made in a greater measure than had been traditionally thought, that in this constructive activity man is not a passive recorder of happenings in the world, that in doing science he uses not only his brain and his senses but also his imagination and decisions. The growth of science is part of the biological process and part of social life. Apart from being an intellectual enterprise guided by the ideal of truth, it is also an activity motivated by the pursuit of beauty (*The Value of Science,* Author's Preface to English translation, p. 8, also Part II, p. 75), and it has an instrumental or pragmatic aspect. It is true that scientific theories which penetrate all facts undergo frequent and radical changes. It is true that mathematicians change formalisms and approaches to their subjects in a way which appears similar to the changes in fashions. To the layman all this seems to amount to the destruction of the traditional view of science as a rational, progressive, accumulative process. And yet, though the traditional view of science certainly requires revision, the fundamental ideals of rationality, objectivity and progress remain intact. Even though conventions which underlie facts and theories do change, yet conventions and changes in conventions are not arbitrary and irrational. Though not imposed uniquely by the structure of our mind—contrary to the belief of rationalists—or by experience—contrary to the empiricist creed, their choice is guided by experience and by other rational considerations such as precision, i.e. reduction of vagueness and the simplifying effect they have on the solution of problems. Many changes in science are merely changes in language as when alternative mathematical formalisms are introduced, or some indifferent hypotheses or theoretical concepts (metaphorical in nature) are replaced by others. Some of the "rival" languages are intertranslatable. Their choice is partly motivated by the prospect of facilitating the solution of problems (in this respect they resemble the choice of coordinates) but partly by the hope of gaining *new points of view* which will "reveal generalizations not suspected before", will reveal "the hidden harmony of things", by making the scientist *"see things in a new way"* (*The Value of Science*, Part II, Chap. V "Analysis and physics", pp. 78–9). Maxwell's achievements in electrodynamics were of this type. They were not due to any new experiment but were rather *a priori* and had to wait twenty years for experimental confirmation (*The Value of Science*, p. 78). The transformations which mathematical and physical theories undergo are often so radical that only "an expert eye finds traces of the prior work of the centuries past" in them (*The Value of Science*, Introduction, p. 14). The invariant traits or the traits that make successive or rival theories commensurate are sometimes to be sought only in some common observational (numerical) consequences, sometimes only in the common abstract structure or even only in functionally equivalent elements. It is true, as epistemological relativists and nominalists say, that many of the conventions in science are more expressive of human psychology and human relations than descriptive of the physical world. However, this is exactly why their replacement by others has no effect on the descriptive content of physical theories and why progress is possible in spite of frequent changes. Scientific knowledge concerns the relations between the phenomena or between the elements of the hidden "profound reality"; it is independent of the transient images or metaphors in which our mind finds it convenient to clothe those relations.

Concluding remarks

It must be emphasised that the aim of this article was *not* to criticise Poincaré's conventionalist philosophy but rather to contribute to its understanding.

Criticisms of the fundamental concept of "convention", of the distinction between conventional and empirical statements, of the distinction between "choice guided by experience" and "experimental testing", of the distinction between a language and a theory have been responsible, as we know, for some of the more recent developments in epistemology, philosophy of language and of science.

One of these criticisms was Quine's campaign against one of the two dogmas of empiricism (rather, against the common dogma of moderate, e.g. Kantian, apriorism and moderate empiricism). Though Quine's critique was directly concerned with the logical-empiricist distinction between analytic and empirical statements (in particular with Carnap's explication and use of this distinction), it applies equally to Poincaré's conventionalism. Indeed, it seems retrospectively surprising both that Poincaré's distinctions were not criticised on behaviourist grounds before Quine, [48] and that Quine did not pay more attention in his critique to Poincaré's writings.

If—as Poincaré insisted—in the course of the development of mathematics and of science some *statements change their status*, for example, some empirical generalisations are elevated to the status of conventional principles and in turn may be demoted for various reasons which may include empirical considerations, then what is the difference between so-called conventional principles and empirical statements? Is not the only difference in the readiness with which scientists are prepared to give them up under the pressure of experience? Similarly, is there any difference—in behavioural terms—between "choice guided by experience" and "experimental testing"?

In reply to these criticisms Poincaré could have used several arguments. First of all, he might point out that Quine's rejection of the analytic/synthetic distinction relies on a strictly behaviourist view of language which itself is highly questionable. Moreover, he would claim, I think, that the distinction between linguistic (conventional) and empirical elements in scientific knowledge is useful even though it may often have to be made artificially or conventionally (which does *not* mean "in a purely arbitrary way"). It is illuminating to see the consequences for epistemological problems of separating in various ways the linguistic (conventional) and empirical components of scientific theories under study. Since we are not normally interested in isolated epistemological problems, considerations of coherence will tend to reduce the role of idiosyncratic criteria of separation. At any rate, if the distinction between "linguistic convention" and "empirical hypothesis" is abandoned altogether, not only is the idea of falsification of theories—so important for scientific rationality—devoid of any clear meaning, but also very little can be said about science and mathematics from the epistemological point of view, which is presumably why Quine wants to "naturalise" epistemology.

Apart from these general considerations, one could defend Poincaré's conventionalist philosophy by drawing on some of the ideas developed later by another conventionalist, viz., Kazimierz Ajdukiewicz.[49] Ajdukiewicz started from a critical analysis of the writings of French conventionalists, in particular of Poincaré and LeRoy, and developed a concept of language and meaning based on the idea of acceptance rules. To characterise a language in this sense—let us refer to it as an "Ajdukiewicz language"—it is necessary to specify not only the lexicon and the rules of syntax but also the meaning rules, i.e., the rules which govern the acceptance of the sentences of the language. *Axiomatic meaning rules* demand an unconditional acceptance of specified sentences (which are therefore called the axioms of the language). For example, to use "not" in the meaning which it has in English (or any other intertranslatable language) one must accept sentences such as "It is not the case that both the earth is round and it is not round"; to use "greater than" in the appropriate sense it has in the language of mathematics one has to accept the sentence "For every x, y if x is greater than y, then y is not greater than x"; similarly, to use "force" in the Newtonian sense one must accept the sentence "$f = m.a$", etc. In contrast to the axiomatic meaning rules, *deductive meaning rules* of an (Ajdukiewicz) language demand the acceptance of a sentence of a specified type, whenever one accepts a sentence of another specified type. For example, if one has accepted the sentence "a is greater than b" one is forced also to accept the sentence "b is not greater than a"; if one accepts "There is no *perpetuum mobile*", then one has also to accept "Every machine in order to work has to have a supply of energy from an external source", etc. Finally, *empirical meaning rules* specify sentences which have to be accepted in the presence of some sentences. For example, if one feels pain one has to accept "It hurts"; if one sees the endpoints of two rods coincide (under normal conditions) one has to accept the sentence "Their length is equal", etc. Empirical meaning rules are present only in empirical languages; formally, however, they may be reduced to axiomatic ones. Since meaning rules in this sense define the relation of consequence on the set of all sentences of a language, every Ajdukiewicz language is an axiomatic (or deductive) system. Essential to this case are, of course, the deductive meaning rules. Indeed an Ajdukiewicz language may be based on the full contingent of all three types of meaning rules but one can also construct languages without axiomatic meaning rules (in which no analytic sentences exist) and—as mentioned already—nonempirical languages have no empirical meaning rules.

Let us look at some of Poincaré's doctrines from this point of view. Poincaré saw metric geometries as merely conventional languages. Indeed they may be seen as languages (in Ajdukiewicz's sense) based on axiomatic and deductive meaning rules only, as long as no coordinative definition of "length" is laid down, or as empirical languages if "length" is governed by an empirical rule. A dictionary which translates one metric geometry into another preserves the deductive relations between the sentences, which is why any of the metric geometries may serve as an instrument for inferences in metric terms. The axioms of a system of metric geometry are, from the point of view under discussion, the axioms of the language; in this sense they are terminological conventions: they have to be accepted as true if the primitive terms which occur in them are to be used in the meaning appropriate to the given system. For example, the parallel postulate has to be accepted (as true) if the term "straight line" is to be understood in the Euclidean sense.

In general, the concept of a linguistic convention (or "analytic sentence", though Poincaré himself used "analytic" only with reference to logical tautologies) can be clarified as follows. The class of linguistic conventions (terminological conventions, analytic sentences, meaning-postulates, conventional principles, etc.) in L is the class of all sentences in L which are governed by the axiomatic meaning rules of L. All sentences of that class have the property that it is impossible to use them properly (i.e. in accordance with the rules of L) without accepting them. The impossibility is logical: the assumption that a speaker of L may use a sentence S governed by an axiomatic rule of L properly and yet not accept S, involves a contradiction.

Though a conventional principle of a language L has thus to be accepted if one wants to speak L, it is possible to abandon a conventional principle by *changing the language*. Moreover, one is forced to do so if a conventional principle in conjunction with other sentences governed by the meaning rules of a language and experiential data yields a contradiction: such a language must be modified or abandoned.

From this point of view one can show—against Poincaré's critics— that it was perfectly possible for him to regard the Principle of Relativity as a conventional statement, which is not falsifiable by any experimental result, and yet to consider giving it up in view of the result of Kaufmann's experiment. The principle, in conjunction with other statements, some of them governed by empirical meaning rules, appeared to yield consequences contradicted by Kaufmann's experiment. Poincaré was prepared to resolve the contradiction by changing the language. Since the Principle of Relativity is equivalent to the statement that the laws of physics are invariant under Lorentz transformations, the transition would have been to another language based on a different group. The choice or change of language (in this case of a group) may be guided—Poincaré insisted—by empirical considerations; this must not be confused with falsification which may affect only empirical hypothesis and may occur only within a fixed language.

One of the points on which Ajdukiewicz's radical conventionalism *seems not* to be consistent with Poincaré's conventionalism, concerns the relation between two languages, one of which has to be abandoned if an inconsistency occurs between statements dictated by the meaning rules of the language. According to Ajdukiewicz, the transition in this case must be to a new language *not* intertranslatable with the old one, otherwise the inconsistency would be inherited by the new language through translation. The transition from classical to relativistic mechanics was, according to Ajdukiewicz, of this type. In general, one of the main tenets of Ajdukiewicz's radical conventionalism was the thesis that there exist languages which share no expressions with identical meaning and whose "closures" are not intertranslatable either. Poincaré, on the other hand, believed that all human languages, sufficiently rich in expressions and based on the same logic, are intertranslatable. The question whether these views of Poincaré and Ajdukiewicz are *indeed* mutually incompatible, is not a simple one and no attempt will be made here to answer it.

Notes

* Read at the graduate seminar of the History and Philosophy of Science Department, Cambridge University, on 26 May 1977. First published in *Studies in History and Philosophy of Science*, 8, 4, (1977).
1. The dates of publication given in the text refer to the first French editions. In the rest of the essay the references will be to English translations (Dover Publications) of the first three books mentioned in the text and to the 1920 French edition of *Dernières Pensées*. It should perhaps be noted that the date of the first French edition of *Science et Méthode* is given in some publications as 1909. For a biography of Poincaré see Gaston Darboux (1914).
2. For the rejection of the Kantian view of mathematics cf. A. Cayley (1883).
3. Poincaré's arguments are given in Poincaré (1908, Chaps. III and IV).
4. H. Poincaré, (1898: t. 6) reprinted in *The Value of Science*, 1905.
5. *Science and Hypothesis* (1902) Part III, Chap. VI, "The classical mechanics"; Chap. VII, "Relative and absolute motion"; Chap. VIII, "Energy and thermodynamics"; Part IV, Chap. X, "The theories of modern physics"; *The Value of*

Science (1905), Part II, "The Physical Sciences"; Part III, "The objective value of science"; *Dernières Pensées*, Chap. II, "L'espace et le temps".

6. H. Poincaré, *Revue de Métaphysique et de Morale* (1902), reprinted in *The Value of Science* (1905) as part 3.
7. P. Duhem, *La théorie physique* (1906), translated as *The Aim and Structure of Physical Theory*, p. 150, footnote 7. Poincaré's claim that there exists an invariant under theoretical change may have been made with Duhem in mind. I could not find a denial of that claim in LeRoy's articles referred to; on the contrary, there is a claim of intertranslatability in one of them.
8. I understand from a private communication that a more detailed discussion of Poincaré's philosophy remained unpublished in the original manuscript of Popper's (1935) which was abridged for publication at the publisher's request.
9. G. Holton (1973), Part II, Chap. 6, Poincaré and relativity, esp. pp. 189–192. Having referred on p. 189 to "an interesting paper" by M. Theo Kahan which credits Poincaré with the view that all laws of geometry and of physics are nothing but useful conventions, i.e., the view which Poincaré explicitly repudiated in his polemic with LeRoy, Holton writes: "I am struck by an apparently quite unrelated, different train in Poincaré, namely that he insisted repeatedly that the relativity principle is simply an *experimental fact*. Thus, in an essay "L'espace et le temps" he wrote "le principe de la relativité physique, nous l'avons dit, est un fait expérimental, au même titre que les propriétés des solides naturels; comme tel, il est susceptible d'une incessant revision..." ["the principle of physical relativity, as we have said, is an experimental fact by the same token as the properties of solids; as such it is susceptible to continual revision"]. Holton then goes on to report how in "La mécanique et l'optique" (reprinted in *Science and Method*, 1908) Poincaré declared the principle of relativity as suspect in view of W. Kaufmann's 1906 experimental results which seemed to disprove Lorentz's (and Einstein's) relativity theory. However, Holton quotes out of context. In "L'espace et le temps" (*Dernières Pensées*, p. 49) Poincaré also discusses the conventional sense of the principle of physical relativity, which serves to define "space". And the very next sentence after the one quoted by Holton reads: "... et la géométrie doit échapper à cette revision; pour cela il faut qu'elle redevienne une convention, que le principe de relativité soit regardé comme une convention..." (*Dernières Pensées*, p. 51). See also *The Value of Science*, Chap. IX, p. 109. In other words, according to Poincaré, firstly, the status of statements in physical theories changes over time and, secondly, many of them have both an empirical and a conventional component.
10. *The Value of Science* (1905, 1958: p. 14).
11. This is, in my view, the gist of Chap. X "The theories of modern physics" of *Science and Hypothesis*.
12. The following is a purely terminological point. Some commentators, having realized that Poincaré's philosophy (of geometry or of science), unlike LeRoy's nominalism, is, after all, not anti-empiricist, think it inappropriate to refer to Poincaré as conventionalist and to his philosophy as conventionalism. However, by the same token, one could argue that it is inappropriate to classify Kant's philosophy as rationalist or apriorist, since it did not deny the existence of synthetic *a posteriori* knowledge. I will use in this article Ajdukiewicz's terminology which distinguishes radical apriorism (represented, for example, by Plato) from moderate apriorism (of Kant), radical empiricism (Mill) from moderate empiri-

cism (Hume, logical empiricists), and similarly extreme conventionalism (nominalism, represented, for example, by LeRoy) from moderate conventionalism of Poincaré (cf. K. Ajdukiewicz: *Problems and Theories of Philosophy*, part 3 (1973)).
13. A Grünbaum, "Carnap's views on the foundations of geometry", in *The Philosophy of Rudolf Carnap* (edited by P. A. Schilpp), (1963); *Geometry and Chronometry in Philosophical Perspective*, (1968); *Philosophical Problems of Space and Time*, 2nd ed., (1973).
14. B. Riemann, *Über die Hypothesen, welche der Geometrie zugrunde liegen*, 1854, published posthumously in *Abhandlungen der Gesellschaft der Wissenschaften zu Göttingen*, Bd. 13. English translation in *A Source Book in Mathematics*, (edited by D. E. SMITH), Dover Publications, New York, 1959.
15. Grünbaum, (1963: p. 673, footnote 142).
16. Grünbaum, (1968: p. 106).
17. Grünbaum, (1968: p. 20).
18. *Science and Hypothesis*, pp. 79, 84 and Chap. XXV.
19. A similar discussion of Poincaré's philosophy may be found in my introductory essay "Radical conventionalism, its background and evolution: Poincaré, LeRoy and Ajdukiewicz" in *K. Ajdukiewicz: The Scientific World-Perspective and Other Essays, 1931–1963*, Dordrecht, 1978.
20. Whenever Poincaré discussed the amorphousness of space he referred to geometrical (mathematical) space, for example in the following passage: "...if by space is understood a mathematical continuum of three dimensions, were it otherwise amorphous, it is the mind which constructs it, but it does not construct it out of nothing..."; (for our mind, Poincaré continues, is guided by empirical considerations). Also Chap. II, "Mathematical magnitude and experiment", in *Science and Hypothesis*, pp. 18–34, where by "the physical continuum"—in contrast to the mathematical—Poincaré apparently means "the continuum" of our sensations and not "physical space" in a realist sense.
21. M. Hesse, "Duhem, Quine and a New Empiricism", in *Can Theories be Refuted? Essays on the Duhem–Quine Thesis* (edited by S. G. HARDING), 1976.
22. *Revue de Métaphysique et de Morale*, **7** (1899).
23. A. Grünbaum, "Carnap's views on the foundations of geometry", in *The Philosophy of Rudolf Carnap* (edited by P. Schilpp), pp. 672–3.
24. Cf. English translation in R. Bonola, *Non-Euclidean Geometry*, Dover, New York, 1955.
25. B. Riemann, "Über die Hypothesen welche der Geometrie zugrunde liegen", in H. Weyl, *Das Kontinuum*.
26. H. Poincaré, "Sur les hypothèses fondementales de la géométrie", *Bull. Soc. Math. Fr.* **XV** (1887) p. 154, as 1877. The year of publication is incorrectly given in Bonola, (1955) p. 154, as 1877.
27. "Tatsachen" also occurs in H. Helmholtz' "Über die Tatsachen die der Geometrie zum Grunde liegen", *Nachr. Königl. Ges. Wiss. zu Göttingen*, 1868, which has given rise to the "Riemann–Helmholtz problem" later solved by S. Lie. However, there is no reference to Helmholtz in Poincaré's 1887 article although there is in (1898a: p. 40). In Lie's *Theorie der Transformationsgruppen*, 1893, it is claimed on p. 437 that in 1887 Poincaré was unfamiliar with either Helmholtz's 1868 or Lie's 1884 work.
28. Riemann, following Gauss, adopted the differential approach to geometry. In this approach one starts from the differential properties (properties in the small) of

surfaces from which the form of the surface and geometric properties as a whole can be subsequently calculated by methods similar to the methods of solving differential equations. Differential geometry was, of course, a suitable instrument for developing field-theories. On the other hand, in his contributions to electrodynamics, Riemann, along with Neumann (1845), Weber (1846) and Clausius (1877), followed the traditional Newtonian model and explained all electrical and magnetic action by means of forces between elementary electrical charges. Cf. M. Born, *Einstein's Theory of Relativity*, pp. 183–327, Dover Publications, New York, 1965.
29. *Proceedings of the Christiania Academy of Science*, Oslo. The main ideas of the memoir appeared in a note by S. Lie and F. Klein in the *Monatsberichte* (15 Dec., 1870) of the Berlin Academy. The English translation of the memoir is in *Source Book in Mathematics*, pp. 485–523 (edited by D. E. Smith).
30. J. Plücker, 1801–68.
31. S. Lie, "On a class of geometric transformations" in *A Source Book in Mathematics* (edited by D. E. Smith).
32. S. Lie, p. 487, M. Chasles and J. Steiner should be mentioned here.
33. Plücker has shown the algebraic sense of the principle of duality: the equation of a straight line in homogeneous coordinates $ax + by + ct = 0$ may be seen either as defining a set of points or as a pencil of lines through the fixed point (x, y, t) if the Cartesian convention is reversed.
34. In *Science and Method* the relevant Chaps are III, "Mathematics and logic", and IV, "The new logics", in particular pp. 171–2, also pp. 182, 184. There is a short but *very unsatisfactory* discussion in *Science and Hypothesis*, Chap. III, "Non-Euclidean geometries: on the nature of axioms", pp. 48–9.
35. The view that the axioms of arithmetic form an implicit definition of arithmetical primitives was held, for example, by Dedekind, Peano, Russell, Couturat and others.
36. *Science and Hypothesis*, Chap. III, "On the nature of axioms", p. 50.
37. S. Lie, "On a class of geometrical transformations"; H. Poincaré, "Sur les hypothèses fondamentales de la géométrie", p. 215; *Science and Hypothesis*, "Lobatchevsky's geometry being susceptible of a concrete interpretation, ceases to be a useless logical exercise, and may be applied. I have no time here to deal with these applications, nor with what Herr Klein and myself have done by using them in the integration of linear equations", p. 43.
38. This result, due to Sophus Lie, is based on research made by Helmholtz and Klein; cf. R. Bonola, *Non-Euclidean Geometry*, pp. 153–4; H. Poincaré, "Sur les hypothèses fondamentales de la géométrie"; Science and Hypothesis, Chap. III, pp. 46–7. Cf. our footnote 28.
39. H. Poincaré, "Sur les hypothèses fondamentales de la géométrie", p. 215; *Science and Hypothesis*, pp. 48–50, 72–3; *Science and Method*, Chap. I. "The relativity of space", esp. pp. 113–16; "Des fondements de la géométrie", pp. 265–9; *The Value of Science*, Chap. III "The notion of space, 5. Space and empiricism", pp 68–9. K. Ajdukiewicz saw the main points of Poincaré's geo-chronometric conventionalism 1. in the insight that the colloquial meanings of metric terms, such as "is congruent" and "is simultaneous", etc., are not sufficient to make statements in which these terms occur decidable by experience, 2. in the insight that the meanings of those metric terms may be made sufficiently precise in several different ways; K. Ajdukiewicz: *Problems and Theories of Philosophy*, pp. 38, 40, (Polish edition) 1949, (English translation) 1973.

40. It is assumed here, as is customary in current discussions, that the Duhem thesis does *not* have strictly nominalist, anti-empiricist implications. Whether such an assumption is justified is another matter and may be seen as dubious in view of Duhem's apparent endorsement of LeRoy's account of science mentioned in the first chapter of the present article. In his discussion of the thesis of the Duhemian interdependence of geometry and physics, attributed to Poincaré by Einstein, Grünbaum considers and rejects the possibility that the Duhem thesis itself might have radical nominalist (hence anti-empiricist) implications due to Duhem's apparent insistence on the ambiguity of all observations. Grünbaum writes: ... For suppose that observations were so ambiguous as to permit us to assume that two solids which *appear* strongly to be chemically *different* are, in fact, chemically identical in all relevant respects. If so rudimentary an observation were thus ambiguous, then no observation could ever possess the required univocity to be incompatible with an observational consequence of a *total theoretical system*. And if that were the case, Duhem could hardly avoid the following conclusion: "observational findings are always so unrestrictedly ambiguous as *not* to permit even the refutation of any given total *theoretical system*". But such a result would be tantamount to the *absurdity that any total theoretical system* can be espoused as true *a priori*... (*Geometry and Chronometry in Philosophical Perspective*, pp. 129–30). However, the apparently absurd conclusion mentioned by Grünbaum does not follow at all *if one refuses to regard any total theoretical system as true or false, which is exactly what an extreme conventionalist or nominalist does*. A theoretical system from a nominalist point of view is neither true nor false not only because it consists of conventional sentences (laws and theories and theory-dependent "observation" sentences) but also because it consists of sentences which are "open-ended" (incomplete "conjunctions", "disjunctions", or "implications") in the sense that they fail to enumerate all relevant conditions (cf. Duhem's view of laws and of observation-sentences).
41. Translation may be based on a dictionary (word by word translation) or in an holistic manner (replacement). It is a characteristic feature of radical conventionalism of Ajdukiewicz that—owing to the claim of the existence of languages which are not intertranslatable and which have no intertranslatable extensions (*closed* not intertranslatable languages)—it affirms the possibility of avoiding falsification by a transition to a language in which the hypothesis in question (or its negation) cannot be expressed at all in its original meaning, cf. *K. Ajdukiewicz*, "Language and meaning", *The Scientific World-Perspective and Other Essays*. 1978; also his "The scientific world-perspective" in *H. Feigl* and *W. Sellars*, *Readings in Philosophical Analysis*, 1949.
42. When Kuhn argued that experience and logic alone do not determine changes in science, he was attacking a view of science unaffected by conventionalist philosophy, cf. *T. Kuhn, The Structure of Scientific Revolutions*, p. 93, 1964.
43. Unless I am mistaken, this point seems to me similar to the view expressed by M. Hesse in her article "Duhem, Quine and a new empiricism", in *Can Theories Be Refuted? Essays on the Duhem–Quine Thesis* (edited by S. G. Harding), 1976.
44. This status of geometries as languages results, according to my account based partly on Poincaré's dispute with LeRoy in *The Value of Science*, Part III, from the "nominalist attitude" which amounts to the convention so to understand the primitive terms of metric geometries that the axioms remain true under the interpretations. However, it must be pointed out that on occasions Poincaré

argued as if, for some obscure reason, it were impossible to give geometrical terms empirical interpretations, i.e. interpretations independent of the axioms. So, for example, at the end of his dispute with Russell in "Des fondements de la géométrie", *Revue de Métaphysique et de Morale*, 7, 279 (1899), Poincaré writes: "... I would ask him [Russell—J. G.] firstly, to explain to me by what specific experiences he would like to verify Euclid's postulate and, secondly, to give me a definition of distance and of straight line which would be independent of that postulate and free from ambiguity and from a vicious circle". It is possible that the intended sense of the challenge in the quoted sentence is the implied claim that the intuitive meaning of "distance" and "straight line" is not precise enough for the empirical testing of geometries, cf. footnote 1 on p. 10. A reference to "ostensive definitions" which give meanings to terms independently of axioms may be found, for example, in *The Value of Science*, p. 45.
45. *The Value of Science*, Chap. VII, "The history of mathematical physics"; also Chap. XII, "Optics and electricity" of *Science and Hypothesis*, the contents of which appeared in *Théorie Mathématique de la lumière* (1889) and in *Electricité et Optique* (1901).
46. *Dernières Pensées*, Chaps. II and III. In Chap. VIII of *Science and Hypothesis* Poincaré discussed the similarities and dissimilarities between geometric and mechanical principles. The main differences which he pointed out are, firstly, that mechanical principles originated more directly from mechanical experiments while geometrical principles are more abstract, secondly, that the conventional part of mechanics if isolated from the empirical part would not be comparable in power and importance with geometry, pp. 137–8.
47. *The Value of Science*, Chap. VII; *Science and Hypothesis*, Chaps X and XII. Observational (numerical-predictive) equivalence and structural similarity (equations of the same form) are sufficient but not necessary conditions for synonymity of theories in Poincaré's sense. Under the influence of Lie's theory of groups of transformations and of evolutionary theory he was inclined to make comparisons between some of the "spaces" constructed by man and by lower animals in the course of their respective behaviour. Some sort of "functional equivalence" would in such contexts be the basis of "synonymity" or at least of comparability (commensurability).
48. The analytic-synthetic distinction had been criticised before Quine but not on the same grounds; for example, Russell in his *An Essay on The Foundations of Geometry* (1897) makes the following remark which one cannot read today without some philosophical melancholy at the coming and going of philosophical modernity or fashions:
"... The doctrine of synthetic and analytic judgements—at any rate if this is taken as the corner-stone of Epistemology—has been so completely rejected by most modern logicians (cf. Bradley's Logic, Bk. III, Chap. VI: Bosanquet's Logic, Bk. I, Chap. I, pp. 97–103), that it would demand little attention here... Every judgement—so modern logic contends—is both synthetic and analytic..." (pp. 57–8).
49. K. Ajdukiewicz, *The Scientific World-Perspective and Other Essays*, 1978.

2

The Physics of the Principles and its Philosophy: Hamilton, Poincaré and Ramsey*

IN AN essay entitled "Theories", written in 1929 (1978), Frank P. Ramsey set out "...to describe a theory as simply a language for discussing the facts a theory is said to explain". Though sketchy, Ramsey's essay is believed by many to contain important insights into the structure and functioning of scientific theories. The most interesting of these appears to be Ramsey's proposal to cast the whole empirical content of a theory in the form of a single sentence (in second order language)—known ever since as "the Ramsey sentence of a theory"—which, in a sense, eliminates all the theory's theoretical (non-observational) terms, replacing them by existentially bound variables.

Ramsey's "Theories" appeared at the time when the philosophy of science was dominated by various extreme forms of empiricism such as Mach's and Russell's phenomenalism, the logical positivism of the early Vienna Circle, Bridgman's operationalism and Watson's behaviourism. All these philosophical schools were anti-metaphysical (in this respect David Hume was their father-figure) and anti-theoretical; they all relied on very narrow–empiricist criteria of meaningfulness to eliminate metaphysical and theoretical (non-observational) terms from science and in this way to reduce science to its empirical–phenomenal basis. Ramsey's analysis of theories was at the time seen in the same light and did not attract much attention. It was not until some time later that Richard Braithwaite, Rudolf Carnap and then others began to see new and interesting features in Ramsey's conception of theories. The first wave of renewed interest in Ramsey came in the fifties and sixties when the empiricist view of scientific theories was examined in connection with the contrast between the programme of the elimination of theoretical terms individually by explicit definitions in observables, and the programme of wholesale replacement with the help of either Ramsey's or William Craig's method (e.g. Braithwaite, 1953; Carnap, 1958, 1966; Hempel, 1958, 1963; Scheffler, 1960, 1969; Nagel, 1961; Bohnert, 1965). The second wave came in the seventies within the context of the discussions over the so-called "structuralist" view of theories (Suppes, 1955, 1957, 1969; Sneed, 1971; Stegmüller, 1976) and the relative merits of the structuralist (or set-theoretic) and the model-theoretic approach to the reconstruction of theories (Przełecki, 1974; Stegmüller, 1976; Tuomela, 1978). Between these two revivals came a widespread and uncompromising attack on the logical empiricist view of science in which, among other things, the very distinction between observables and theoretical terms was rejected and the scientific and philosophical importance of theories (understood in a realist way) and of underlying metaphysical systems defended or even extolled (Popper, 1956, 1963, 1972; Putnam, 1956, 1979; Polanyi, 1958; Toulmin, 1961; Kuhn, 1962; Feyerabend, 1964; Lakatos, 1970; Holton, 1973). In the eyes of the critics just mentioned (Popper, in particular) Ramsey's conception is a variety of the instrumentalist view of theories, according to which only observables have a descriptive (cognitive) meaning whereas theoretical terms are mere formal, non-descriptive symbols. It is thus a variety of phenomenalism or of positivism which they condemn and reject on several grounds.

No attempts have been made so far to trace back the historical origins or at least antecedents of Ramsey's view of theories, except that it is often linked with the philosophy of Pierre Duhem whether by the critics of that philosophy (e.g. Popper, 1963, Chap. 3) or by those who emphasise its merits (e.g. Sneed, 1971). However, Duhem's philosophy of science is complicated. Some of Duhem's writings appear to support unambiguously the claim that it is a paradigm of instrumentalism as briefly characterised at the end of the previous paragraph. Others, however, seem to contradict this interpretation (for example, the idea of "natural classification", and the doctrine of the theory-dependence of all observables). Without denying that there is strong affinity with some of Duhem's ideas, the historical origin or antecedence of Ramsey's conception of theories will be traced in the present essay to what Henri Poincaré in (1905) called "the physics of the principles" as it may be found in the works of Lagrange, Hamilton, Fourier and others. More specifically Hamilton's method in geometrical optics will be discussed as an example of a replacement method in the sense akin to Ramsey's. If we decide to extend the term "instrumentalism" to any (part of a) philosophy which approves of replacement procedures for some purpose and under some circumstances at least, then the following features of instrumentalism must be taken into account to revise its image. In the physics of the principles the replacement method relies on abstract, often sophisticated mathematical principles which are used to summarise the empirical laws and experimental facts common to two or more rival theories. The language of the principles is not purely phenomenal for it involves quite advanced, abstract mathematics. This is, however, not due to a mere linguistic preference but rather to the epistemology (often apriorist; Hamilton, for example, was an orthodox Kantian) according to which our mind plays an active role in cognition and this is exercised in science mainly through the good offices of mathematical concepts, principles, formal analogies, etc. The replacement procedures are not guided by the empiricist criteria of what is meaningful and are not designed to eliminate all theoretical terms from science. There is, therefore, no need for a clear and sharp distinction between observables and theoretical concepts across the whole language of science. The intention is rather to identify what appear to be intractable theoretical disputes or rivalries which have defied for a long time experimental resolutions, such as that between the emission (particle) and the undulatory (wave) theories of light; to extract their common empirical content and common structure (hence concepts, *prima facie,* common to rival theories are classified as observational) in terms of general principles and in this way to raise oneself above the theoretical dispute in order either to continue research within the common ground or else to allow free choice of either of the theoretical frameworks (the phrase "to raise an empirical law or a hypothesis to the status of a principle", popularised by

Poincaré's conventionalist philosophy, derives from this context). Apart from pragmatic considerations concerned with scientific research strategy, the physics of the principles and its replacement methods have a deeper epistemological motive. It is the conviction, expressed most emphatically in Poincaré's epistemology, that there are definite limits to theoretical knowledge, limits which consist in the observational equivalence and formal similarity of alternative ("rival") theories, or—if we prefer to put it in ontological rather than linguistic terms—in the indistinguishability (duality reciprocity) of "worlds" (spaces) which can be transformed one into another by any continuous point transformation. Instrumentalism (in this sense) is not concerned mainly with what is meaningful and what meaningless in our language but rather with what can and what cannot be known within science. The answer to the latter question cannot be given on the basis of "pure philosophical reflection" but rather on the basis of a critical examination of actual scientific progress. Negative results of attempts to detect experimentally absolute motion or the motion of bodies relative to the ether do not on their own support the semantical claim that "absolute motion" (or "velocity relative to the ether") is a meaningless term (its meaning can be specified in terms of possible positive results) but rather that absolute motion, even if it should exist, cannot be known. As a matter of historical fact, the dispute concerning absolute and relative motion had been seen by "mathematicians" since Newton and Leibniz in this epistemological aspect (Can absolute motion be known?). The linguistic paraphrase came from phenomenalist philosophers (Berkeley, Mach).

There is no evidence that Ramsey had the physics of the principles in mind when he wrote his article on theories or that he was even familiar with Hamilton's method in geometrical optics. However, Cambridge in Ramsey's time was one of the most likely places where such an association of ideas would occur. It was there at the turn of the century that Edmund T. Whittaker's lectures and his book on dynamics (1904, esp. Chaps. IX through to and including XII) had anticipated what some twenty years later was to become "the Hamiltonian revival". Although in 1906 Whittaker left Cambridge to take up appointments as Royal Astronomer for Ireland and Andrews

Professor of Astronomy in the University of Dublin—the same posts which once had been held by Sir William R. Hamilton—several of his pupils remained at Cambridge (see McCrea, 1957; Temple, 1956). In quite a few areas of research Whittaker may be seen as Poincaré's successor. This applies especially to Whittaker's works on automorphic functions (1898, 1929, 1930) and on the integration of Hamiltonian equations (1902, 1917). One of the most distinctive features of his *Analytical Dynamics* (1904) is the presentation of the transformation theory of dynamics (Chapter XI) which links it closely with Lie's and Poincaré's quasi-geometrical approach to physics.

Another Cambridge scientist whose views on theories are likely to have influenced Ramsey's philosophy, and have definitely influenced Poincaré's, was Sir Joseph Larmor. In "A dynamical theory of the electric and luminiferous medium" (1894) Larmor argued that the problem of devising a theory concerned with "the determination of the constitution of a partly concealed dynamical system, such as the ether" can be divided into two independent parts: the first part is the determination of some form of energy-function which would describe the recognised dynamical properties of the system and could be tested by further applications; the second part is the construction (in actuality or imagination) of a mechanical model or illustration of a medium possessing such an energy-function (p. 721). Failure to construct a model should not, however, be seen as discrediting the solution already achieved of the first part of the problem. In the same article, as well as in *Aether and Matter* (1900, 1972: pp. 213–14), Larmor emphasised the already well-established method of enunciating any dynamical problem as a variational one, in a single formula, "... in order to help in the reduction to dynamics of physical theories in which the intimate dynamical machinery is more or less hidden from direct inspection". If the laws of any department of physics can be formulated as a minimal or variational principle, the subject is thereby effectively reduced to the dynamical type and what remains are interpretations, explanations and analogies which correlate the integral that is the subject of variation with the corresponding integrals relating to known dynamical systems.

In the (1894) article Larmor suggests that one can trace the formal relations between alternative theories of physical optics and of electricity by comparing them with "the electromagnetic scheme of Maxwell" (1894: p. 735). This was done by Poincaré in "À propos de la théorie de M. Larmor" (1895) (see Chap. 5, this volume) and became subsequently one of the bases for his conventionalist analysis of the status and role of theories in 1902 Chap. X).

Ramsey on Theories

In "Theories" Ramsey constructed a simple example of a theory with two "universes of discourse", one called the primary and the other the secondary system. Propositions of the primary system, which represent facts to be explained, are truth-functions of quantifier-free expressions formed from predicate or function symbols A, B, C, D, etc., and a suitable number of individual constants as names of the individuals of the primary system. These individuals might be, for example, instants of time, in which case their names would be integers (or rationals) and the propositions would be one-valued numerical functions, e.g. $g(3) = 1$. Such atomic propositions and at least some of the truth-functions of such atomic propositions are empirical in nature.

The secondary system is obtained by a single expansion of the primary system. New predicate or function symbols α, β, γ, etc., are introduced (which can be thought of as the theoretical or abstract terms of the theory) to form truth-functional propositions with the individual constants of the primary system, $\alpha(n)$, $\beta(n)$, $\gamma(m, n)$, etc. These are constrained by a finite number of axioms, i.e. formulae whose predicate or function terms belong exclusively to the vocabulary of the secondary system and whose quantifiers range over the universe of discourse of the primary system, e.g. $(n)\ \alpha(n) \cdot \beta(n)$. Whatever formulae of the secondary system are deducible from these axioms (presumably on the basis of first order classical logic) are called theorems. The two systems are then linked by a dictionary, i.e. a set of equivalence definitions of the predicate or function symbols of the primary system in terms of the symbols of the secondary system. General propositions in A, B, C, D, etc., deducible from the conjunction of the axioms and the dictionary are called "laws" and represent empirical generalisations. Singular propositions in *A, B, C, D*, etc., deducible from that conjunction are called "consequences". The totality of laws and consequences, i.e. of universal and singular propositions in which no extralogical symbols of the secondary system occur and which are deducible from the axioms in conjunction with the dictionary form the eliminant and it is only this eliminant-totality that the theory asserts as true (1978: p. 104).

Having made the above distinctions, Ramsey considers six questions about theories, the first four of which seem to be most important for the understanding of his views on theories. The four questions are as follows.

1. Can we say anything in the language of a theory (i.e. of the secondary system) that we could not say without it?

2. Can we reproduce the structure of a theory (i.e. the secondary system) by means of explicit definitions within the primary system?

3. Is this necessary for the legitimate use of the theory (as Russell, Nicod and Carnap insisted)?

4. If not, how are we to explain the functioning of a theory without such definitions? (Ramsey, 1978: pp. 108–9, 112). He answered these four questions in this way.

1'. Obviously not, since we can easily eliminate all the specific terms of the secondary system and so say in the primary system "all that the theory gives us".

2'. There is no simple way of inverting the dictionary, i.e. of defining the non-logical terms of the secondary system using only the expressions of the primary system; nor is it possible to express each proposition of the secondary system in terms of the necessary and sufficient conditions formulated in the language of the primary system; however, it *is* always possible to define explicitly non-logical terms of the secondary system in terms of the primary system provided that definitions of any complexity are allowed *or* the universe of discourse is finite.

3'. Although specific terms of the secondary system are explicitly definable in terms of the primary system (with the proviso given in (2')), it is not necessary for the legitimate use of a theory actually to provide such definitions.

4'. The functioning of a theory without such definitions may be best explained if the original theory, which we assume to be given as a (finite) conjunction of the axioms and the "dictionary" both in first order language, is rewritten in the form of a sentence obtained by second order existential generalisation on all extra-logical terms occurring in the axioms and in the dictionary. To use Ramsey's informal abbreviations (slightly modified), we rewrite the original theory in the form of the sentence (the Ramsey-sentence, RS):

$$(\exists \alpha', \beta', \gamma'): \text{dictionary.axioms} \qquad \text{(RS)}$$

where α', β', γ' are (second order) variables replacing corresponding extralogical constants (α, β, γ) of the secondary system; "dictionary" and "axioms" are sentential forms obtained, respectively, from the dictionary and the axioms by writing suitable variables in place of corresponding extra-logical constants (1978: p. 120). A theory so reformulated is to be understood in such a way that, firstly, what it asserts is only the conjunction of its "laws and consequences" (both in the language of the primary system), secondly, the expressions of the secondary system are "simply a language" in terms of which the laws and consequences may be "clothed"; otherwise sentences in which the terms of the secondary system occur are not "strictly propositions by themselves", i.e. when isolated from the context of the theory, just as the sentences in a story beginning "Once upon a time" have no complete meanings and so are not propositions by themselves (1978: p. 120). This does not affect at all our *reasoning* but may affect our *disputes* (1978: pp. 121–2). In other words, the deductive elaboration of a theory is not affected by the incomplete nature of isolated sentences of the secondary system; moreover, two theories which differ with respect to their secondary systems need not contradict one another (1978: p. 122), though they may appear to do so. Finally, the Ramsey-sentence formulation of a theory makes it clear that questions about the meaning and truth of sentences of the secondary system may only be answered in the context of the whole theory and not in isolation from it (1978: pp. 120–1).

Apart from the four questions, 1. through 4., Ramsey raises the problem of intertheory relations, viz. the problem of the clarification of the meaning of "two contradictory theories", "two equivalent theories", "one theory contained in (reducible to) another". He sketched definitions of these phrases in terms of "content of a theory", which—as we have seen—he identified with "the totality of laws and consequences of a theory". So, for example, for two theories to be equivalent it is necessary and sufficient that their content, i.e. sets of laws and consequences be equivalent (1978: pp. 121-2). Although he did distinguish between the content of a theory (what the theory asserts) and its symbolic form and pointed out that given two equivalent theories "there may be more or less resemblance between their symbolic form", nevertheless he concluded that this type of resemblance is difficult if not impossible to define precisely (1978: p. 122).

Duhem's Instrumentalist Interpretation of "Saving the Phenomena"

Whether Ramsey was aware of it or not, views concerning the structure and functioning of scientific theories similar to his, in some important respects at least, had a long tradition in science and its philosophy. While this fact may moderate somewhat our appreciation of his originality, it may also indicate that his account of scientific theories is not a product of a transient philosophical fashion.

Pierre Duhem is often mentioned today as one of those who anticipated the view that a scientific theory consists of two integrated systems: the primary system expressing experimental facts and laws and the secondary system which is a mathematical, uninterpreted calculus (Abraham and Marsden, 1967: p. 231). He is also credited as one of the fathers of the holistic view of theories since he insisted firstly, that experimental laws take different meaning depending on the theory into which they have been absorbed; secondly, that only whole theoretical systems can be tested against facts but not isolated hypotheses (Duhem, 1906, 1969, Part II, Chaps. V, VI). For these reasons his name is sometimes explicitly linked with Ramsey's (Sneed, 1971: Chap. IX). (For Duhem's view on mechanical models see Hesse (1963)).

We know, however, that Duhem's philosophy of science was based on his research into the history of astronomy and physics "from Plato to Galileo", the results of which he summarised in Duhem (1906, 1969). According to Duhem, an adequate, critical account of the structure and functioning of scientific theories emerged gradually from the writings of Greek astronomers and cosmologists. Plato, who on *a priori* grounds believed that the motion of the heavenly bodies is circular, uniform and regular (in the same direction), set mathematical astronomers the task of finding hypotheses which would "save the phenomena presented by the planets" in terms of circular, uniform and perfectly regular motions. Henceforth ancient astronomy was concerned with geometrical constructions (mathematical models in modern terminology) which would assign each planet a path conformable to its visible path. Hipparchus appears to be one of the first to have discovered and proved that the same phenomena, for example the observed course of the sun, may be equally well accounted for by two apparently quite different hypotheses, viz. that of eccentric circles and that of concentric circles bearing epicycles. The two hypotheses were subsequently shown to be observationally equivalent, which fact persuaded Theon to claim that astronomy is not able ever to discover *the true* hypothesis and that the disputes among astronomers concerning apparently different but observationally equivalent hypotheses may be dismissed as idle. Some authors, Geminus, for example (as reported by Simplicius), drew from these circumstances the conclusion, which appeared to conform to Aristotle's teaching, that the aims and methods of mathematical astronomy were different from those of physics, concerned with the discovery of the nature and causes of things, and that astronomers needed the help of physical principles to choose the true hypotheses from the totality of all observationally adequate ones. Others (Adrastus, Theon, Dercyllides) proclaimed that an astronomical hypothesis cannot be true to the nature of things unless craftsmen are able to construct its representation in hardware (wood or metal). These suggestions were rejected by Claudius Ptolemy who—according to Duhem—concluded that, firstly, an adequate astronomical theory is a geometrical contrivance which saves all observed astronomical phenomena; secondly, astronomers are unable and do not have to discriminate between observationally equivalent hypotheses (which save all the phenomena equally well); thirdly, they therefore do not and must not attribute physical reality to the conceptual elements in their models (Duhem, 1969: pp. 5–20). These views were adopted subsequently by Proclus who, moreover, gave the following logical analysis of the method of astronomy:

> (astronomers) do not arrive at conclusions by starting from hypotheses, as is done in the other sciences; rather, taking the conclusions as their point of departure, they strive to construct hypotheses from which effects conformable to the original conclusions follow with necessity (Duhem, 1969; p. 20).

Proclus's account of the method of astronomy seems to be equivalent to Charles Peirce's description of the method of abduction (inverse induction) to which other authors give the name of "reduction" or "reductive inference" (Łukasiewicz, 1912, 1970: pp. 6–7) or "the method of hypotheses" (Popper, 1963). Taken in itself, Proclus's description of the method of astronomy is ambiguous with respect to the role of secondary (abstract, theoretical) system and therefore may and has been used by both instrumentalists and realists (in one of the senses of these words), viz. by those who see in the secondary system an empirically uninterpreted calculus (as Ramsey did) and those who attribute to it a descriptive function. But even in conjunction with the reported three theses of Ptolemy's (who, by the way, did not consistently hold them) it would only amount to a "formalist" or "instrumentalist" view of the aims and method of astronomy, not of all science. According to Duhem, *the generalisation of the instrumentalist view* of the aim and nature of astronomical theories to all theories of physics occurred later in the writings of mediaeval and Renaissance philosopher-scientists to be confronted by the similarly general realism of Kepler and Galileo. It is worth noting, however, that Duhem's reading of the history of science tended to emphasise the importance of the two extreme doctrines of (generalised) instrumentalism and realism, as if either were the limit of one of two rival, convergent series of historically evolving doctrines. An alternative reading would rather stress the existence and legitimacy of a variety of intermediate views, with varying amount of instrumentalist and realist elements. One class of these intermediate doctrines may be called restricted instrumentalism, and one specific doctrine in this class would be the view that the theoretical postulates of a certain scientific discipline practised in isolation from others and, therefore, in abstraction from some relevant factors, may only be accorded the status of empirically uninterpreted sentences. Such restricted instrumentalism was seen as advantageous by many ancient astronomers who preferred to carry on their research as "mathematical astronomers" independently of (Aristotelian) physics.

Hamilton's Method in Geometrical Optics

Ancient and early modern astronomers used ordinary Euclidean geometry as the language of their "secondary systems". The discovery of coordinate and analytic geometry by René Descartes and Pierre Fermat, the development of the differential and integral calculus by Isaac Newton and Gottfried Leibniz and of the calculus of variations by Joseph Lagrange, had the effect that analysis became the language of modern theoretical physics. Henceforth the concept of a mathematical model became synonymous with the idea of a set of differential equations, except where probability distributions enter the picture. A set of differential equations in such a model is usually expressible equivalently in the form of an extremal (stationary) principle. All one has to do is to find or construct a quantity which—when subjected to some constraints—takes on an *extremum* value.

The theory of mathematical (geometrical) optics developed between 1827 and 1832 by Sir William Rowan Hamilton (1805–1865) is an example of a theory on which Ramsey's description of theories could have been modelled except that Hamilton's starting-point—unlike Ramsey's—were *two* apparently rival physical theories, viz. the emission and the wave theory of light restricted to a common area of application (provided that the interference phenomena are disregarded) and the mathematics sufficient to systematise known experimental results in that area. Hamilton's method seems to have achieved for the overlapping parts of two rival theories what the Ramsey sentence does for one theory. It must be emphasised that far from being a trivial and sterile exercise, Hamilton's approach to geometrical optics proved to be very fruitful. It led in connection with his study of Fresnel's surface to the theoretical prediction of a new, previously unknown phenomenon, viz. conical refraction, subsequently verified by Humphrey Lloyd's experiments (Hamilton, 1833a, 1931: pp. 302-3), and to the discovery of a new form of equations of dynamics (Hamilton's wave mechanics) (Hamilton, 1833b, 1834a, b).[1]

From Hamilton's own accounts we know that he was prompted to undertake his theoretical research in optics by the following considerations. In spite of an abundance of experimental laws concerning optical phenomena accumulated over the centuries by ancient and modern physicists, there did not exist a mathematical theory which could systematise them and which would be comparable in precision, practical usefulness and formal beauty with other well-developed branches of science such as Lagrangean mechanics (Hamilton, 1833b, 1931: p. 315). Two rival theories of light, Newton's emission theory and Huyghens's wave theory, competed for the attention of scientists and each of them in turn appeared for a time to win the upper hand. Two rival general principles, that of least action (Maupertuis's law) and that of least time (Fermat's law), each associated with one of the two rival theories, were moreover originally understood in metaphysical terms as expressing teleology (goal-directedness), simplicity or economy of Nature. Yet whether one adopts the Newtonian (emission) or the Huyghensian (wave) theory of light to explain the experimental laws of optics, one may regard the linear paths of light, their properties and relations between them, i.e. systems of rays and experimental laws concerning them as the object of an important separate science, viz. mathematical optics (Hamilton, 1833b, 1931: p. 314).[2] The general problem which Hamilton proposed to himself in that area was to investigate the mathematical consequences of the (optical) law of least action as a possible general principle on which to base mathematical optics, "without adopting either the metaphysical or (in optics) the physical opinions that first suggested the name ('action' and 'the law of least action')" (Hamilton, 1833b, 1931: p. 318). The method which he decided to use for that purpose was suggested to him by René

Descartes's algebraic (analytic) geometry. A system of optical rays, i.e. a combination of straight or curved paths along which light is supposed to be propagated according to the law of least action, is characterised in this new approach by one single "characteristic relation". All the mathematical properties of the given system of optical rays are deducible from this characteristic relation in the same manner as the properties of a curve or of a surface are deducible in Cartesian analytic geometry from appropriate functions (Hamilton, 1831, 1931: p. 295). Whereas in Cartesian geometry two or three elements are involved in a characteristic relation, viz. the elements of position of a variable point which has for *locus* a curve or a surface, the elements involved in a characteristic relation of a system of optical rays are eight in number: six elements of position of two variable points in space visually connected, i.e. final and initial coordinates, an index of colour and *action* (or time) between the two variable points. The last one, viz. action, reflects through its dependence on the preceding seven all the geometric properties of the given system of rays, hence it is referred to as the *characteristic function* of the system. Denoted by V, the characteristic function is the integral given by the formula:

$$V = \int v ds$$

where v is the medium function (refractive index of the medium) and ds is the element of the path (ray).

The method based on the characteristic function V of studying optical phenomena was described by Hamilton (1837, 1931: pp. 168–9) in the following way[3]:

Assume that light is propagated according to the optical law of least action (or of swiftest propagation) along any curved or polygon ray, describing each element of the ray

$$ds = \sqrt{(dx^2 + dy^2 + dz^2)}$$

with a molecular velocity or undulatory slowness v. The latter (the medium function) depends in general on the nature of the medium, the position and direction of the element and the colour of the light, in other words, v is a function of the three rectangular coordinates x, y, z, the three cosines of direction (cosines of the angles which the element of the ray makes with the axes of coordinates) α, β, γ:

$$\alpha = \frac{dx}{ds}, \quad \beta = \frac{dy}{ds}, \quad \gamma = \frac{dz}{ds},$$

and a chromatic index χ. Let us denote as follows the variation of v:

$$\delta v = \frac{\delta v}{\delta x}\delta x + \frac{\delta v}{\delta y}\delta y + \frac{\delta v}{\delta z}\delta z + \frac{\delta v}{\delta \alpha}\delta \alpha + \frac{\delta v}{\delta \beta}\delta \beta + \frac{\delta v}{\delta \gamma}\delta \gamma + \frac{\delta v}{\delta \chi}\delta \chi,$$

and, further, in view of $\alpha^2 + \beta^2 + \gamma^2 = 1$, let us determine

$$\frac{\delta v}{\delta \alpha}, \quad \frac{\delta v}{\delta \beta}, \quad \frac{\delta v}{\delta \gamma},$$

i.e. the partial differential coefficients of v, so as to satisfy the condition

$$\alpha \frac{\delta v}{\delta \alpha} + \beta \frac{\delta v}{\delta \beta} + \gamma \frac{\delta v}{\delta \gamma} = v,$$

by making v homogeneous of the first degree with respect to α, β, γ.

On these assumptions it has been shown (Hamilton, 1830, 1931: pp. 107–44) that the variation of the definite integral

$$V = \int v\, ds$$

considered as a function (the characteristic function) of the final and initial coordinates, i.e. *the variation of action or the time* expended by light of any colour in going from one variable point to another is given by the following formula (variational principle):

$$\delta V = \left(\delta \int v\, ds =\right) \frac{\delta v}{\delta \alpha}\delta x - \frac{\delta v'}{\delta \alpha'}\delta x' + \frac{\delta v}{\delta \beta}\delta y - \frac{\delta v'}{\delta \beta'}\delta y' + \frac{\delta v}{\delta \gamma}\delta z - \frac{\delta v'}{\delta \gamma'}\delta z':$$

(A)

where the accented quantities are the initial ones. Formula (A) may be resolved into the following six equations:

$$\left.\begin{array}{lll}\dfrac{\delta V}{\delta x}=\dfrac{\delta v}{\delta \alpha}; & \dfrac{\delta V}{\delta y}=\dfrac{\delta v}{\delta \beta}; & \dfrac{\delta V}{\delta z}=\dfrac{\delta v}{\delta \gamma}; \\ -\dfrac{\delta V}{\delta x'}=\dfrac{\delta v'}{\delta \alpha'}; & -\dfrac{\delta V}{\delta y'}=\dfrac{\delta v'}{\delta \beta'}; & -\dfrac{\delta V}{\delta z'}=\dfrac{\delta v'}{\delta \gamma'}\end{array}\right\} \quad (B)$$

If we eliminate from (B) the ratios

$$\frac{\alpha}{\gamma}, \frac{\beta}{\gamma}, \frac{\alpha'}{\gamma'}, \frac{\beta'}{\gamma'}$$

of which the partial derivatives of v are functions, then we obtain two partial differential equations of the first order between the characteristic function V, the coordinates and the chromatic index (colour):

$$\left.\begin{array}{l}0=\Omega\left(\dfrac{\delta V}{\delta x}, \dfrac{\delta V}{\delta y}, \dfrac{\delta V}{\delta z}, x, y, z, \chi\right) \\ 0=\Omega'\left\{-\dfrac{\delta V}{\delta x'}, -\dfrac{\delta V}{\delta y'}, -\dfrac{\delta V}{\delta z'}, x', y', z', \chi\right)\end{array}\right\} \quad (C)$$

Let us now recapitulate in words the significance which Hamilton associated with his method in optics.

All the problems of mathematical optics are reducible to the study of the characteristic function V with the help of the variational principle (A) (Hamilton, 1831, 1837; 1931: pp. 169, 295).

Hamilton referred to the principle (A) variously as "the principle of least (constant) action" or "the principle of swiftest propagation", etc., but sometimes he designated it "by the less hypothetical name of the Equation of the Characteristic Function" (Hamilton 1828, 1830, 1833a, b, 1837; 1931: pp. 10, 107–9, 168, 297, 311–18). The last phrase was used by him to emphasise that mathematical optics is logically independent either of the wave (undulatory) or of the particle (emission) theory of light. In fact, as indicated before, formula (A) may be transformed either into the principle of least (stationary) action of the emission theory (the Maupertuis principle) or into the principle of least time (the Fermat principle) of the wave theory by suitable interpretation of v in terms of the velocity of the ray element on the assumption of the emission theory or in terms of the "Undulatory slowness" if one assumes the wave theory. Accordingly V will be either the action integral or the time integral.

Assuming the emission theory of light, we see from equations (B) that

> the coefficients of the variations of the final coordinates in the variation of the integral called action are equal to the coefficients of the variations of the cosines of the angles which the element of the ray makes with the axes of coordinates in the variation of a certain homogeneous function of those cosines this ... function ... being equal ... to the velocity of the element ... estimated on the hypothesis of emission ... (Hamilton, 1830, 1931: p. 107).

However, formula (A) "gives immediately the differential equation of that important class of surfaces, which on the hypothesis of undulation are called *waves*, and which on the hypothesis of molecular emission may be named *surfaces of constant action* ... " (1830, 1931: p. 107).

Assuming the wave theory of light, the partial derivatives of V with respect to final coordinates in equations (B) represent the components of "normal slowness of propagation of a wave", the function V represents the time of the propagation of light from the initial to the final point and waves are represented by the general equation

$$V = \text{const.} \qquad (C)$$

On this understanding equations (B) state that partial derivatives of V with respect to final coordinates, i.e. the components of "normal slowness of propagation of a wave" are proportional to the direction-cosines of the normal to the wave for which the time V is constant (Hamilton, 1837, 1931: p. 277; 1833b, 1931: p. 329).

In view of the mentioned correspondence or duality between the two interpretations one may say that "in geometrical optics it is possible to regard the two theories (the corpuscular and wave theory) as different aspects of a single theory" (Synge, 1937: p. 12). Put another way, the ray theory and the wave theory are reconciled in geometrical optics at least so far as ordinary media are concerned (Synge, 1937: p. 16). This fact had a great heuristic value since it suggested to Hamilton the idea of wave mechanics (for Hamilton's own account see Hamilton, 1834, 1940: pp. 212–16):

> The basic feature of Hamilton's method in optics is the reconciliation of a minimal principle (Least Action, or Fermat Principle) with a contact transformation (construction of Huyghens) and this fundamental duality is carried over into dynamics, in which field Hamilton, starting from the well-known principle of Least Action, reduced the equations of motion to an infinitesimal contact transformation... (Conway and Synge, 1931: p. 487).[4]

Instead of "reconciliation" and "duality" we may prefer to use the terminology of "observational equivalence" and "translation", especially if we want to link Hamilton's method in optics with the Ramsey view of theories. Since from the variational principle (A) all the experimental laws of geometrical optics (laws of reflection, refraction, the theorem of Malus, etc.) are derivable and since these laws form the common empirical content of the two theories of light, the latter two theories (wave and particle theory of light) are observationally equivalent within geometrical optics. We can say, therefore, that the particle and the wave theory of light are merely two different but intertranslatable languages so far as geometrical optics is concerned, or that they are two different but equivalent ways of talking about the same things. Systems of optical rays in Hamilton's terminology (Hamilton, 1828, 1931: pp. 1–106) as combinations of visual lines of light are models (realisations) of the theory of geometrical optics and thus models of the common observational part of the particle and wave theory of light (when certain phenomena such as interference are disregarded). These models can be extended to the models of either the particle or the wave theory of light and a correspondence established between these extended models. The variational principle (A), understood in Hamilton's neutral or "less hypothetical" way, asserts that *there exists* a quantity, viz. the characteristic function V which has a stationary property and which is such that all the properties of systems of optical rays—in effect all experimental laws of geometrical optics—are derivable from it. The principle can, therefore, be regarded as the Ramsey sentence formulation of either the particle or of the wave theory of light restricted to geometrical optics. It can be said to *replace* either of these theories within geometrical optics and, in this sense, to eliminate from geometrical optics the theoretical terms "particle of light", "light wave", and others associated with them. Alternatively, one may say that the two sets of theoretical terms and corresponding principles have been shown by Hamilton's method to be *interchangeable* within geometrical optics: the characteristic function V whose existence is asserted by the principle (A) is identified with the action integral on the particle interpretation and with the time integral on the wave interpretation. One is thus left to choose whichever theory one prefers and finds heuristically more valuable.

Having argued that there is a close affinity, logically speaking, between Hamilton's principle in geometrical optics and the Ramsey-sentence formulation of a theory as well as between Ramsey's view of the functions of a theory on one hand and the motives which guided Hamilton in devising his method in geometrical optics on the other hand, are we to conclude that Hamilton's views on the epistemological status of theories were the same as Ramsey's?

In order to answer this question we ought to consider the following facts. First of all, there is no clear indication in Hamilton's writings that he subscribed to a view equivalent to Ramsey's claim that theoretical terms (i.e. terms of the secondary system) are best seen within the context of a scientific theory as empirically uninterpreted symbols or that these terms should be in principle eliminable on the basis of explicit definitions or otherwise. In geometrical optics he believed that a mathematical theory neutral with respect to the wave-particle controversy would be beneficial at the time, especially if one utilised heuristic suggestions associated with the images, analogies, etc., implicit in either of these two theories. But he *did* consider and advance arguments in favour of Fresnel's theory in physical optics. Was it then a case of what we have labelled "restricted instrumentalism"? Before we commit ourselves to this solution, one further point should be mentioned. Like many of his great contemporaries among mathematicians Hamilton *did* have philosophical views concerning science and mathematics and he *did* write about them, however briefly and parenthetically. For our present purposes his views which concern the epistemological status of algebra and analysis are relevant because they appear to be in contradiction with an instrumentalist or formalist epistemology. Hamilton's philosophy of algebra and analysis may be found in his paper on algebraic couples, which also includes an essay on "algebra as the science of pure time" conceived in the spirit of Kantian philosophy (Hamilton, 1833c, 1967: pp. 3–7). In the introductory remarks to that paper Hamilton explained the philosophical motivation behind his research in algebra and analysis which eventually led him to the invention and development of the theory of quaternions. Algebra—by which Hamilton meant both algebra and analysis in our sense (Hamilton, 1833c, 1967: p. 6, footnote)—may be studied in three different ways, viz. as a practical art, as a symbolic language and as a science. Accordingly there emerged three schools of thought, the practical, the philological and the theoretical, each favouring a corresponding approach. However, although algebra (*cum* analysis) proved to be a very useful art and a beautiful language, the state of its theoretical principles—unlike the state of the principles of Euclidean geometry (Hamilton was not familiar with the discovery of non-Euclidean geometries)—left much to be desired:

> No candid and intelligent person can doubt the truth of the chief properties of *Parallel Lines*, as set forth by Euclid in his *Elements*, two thousand years ago.... But it requires no peculiar scepticism to doubt, or even to disbelieve, the doctrine of Negatives and Imaginaries, when set forth ... (on principles used in Hamilton's time—J.G.) (Hamilton, 1833c, 1967: p. 4).

This is why a tendency could be perceived to the rejection of that view which regarded algebra "as a science in some sense analogous to geometry" and to the adoption of the view of algebra as an art or as a language, as a system of rules or else as a system of expressions, but not as a system of truths. One tends thus to substitute for the theoretical question "Is a theorem of algebra true?", the practical question "Can it be applied as an instrument, to do or to discover something else in some research which is not algebraical?" or else the philological question "Does its expression harmonise, according to the laws of language, with other algebraical expressions?" (Hamilton, 1833c, 1967: pp. 4–5). Hamilton mentions George Peacock as the leading representative of the philological view of algebra (in the narrow sense) in Britain and Martin Ohm in Germany. Lagrange is mentioned by him as representing the philological view of the calculus. It was in order to counter that "philological" or formalist tendency that Hamilton undertook the task of laying down the theoretical principles of algebra and analysis as *the science of variation and progression*—hence in his Kantian view—as the science based on the *pure intuition of time*. The theory of algebraic couples (intended to replace the then current theory of negatives and imaginaries), of algebraic triplets and, finally, of quaternions, were the result of his research. Within the theory of quaternions Hamilton developed and used the idea of a 4-space, one dimension of which represented time (one real axis represented time, three imaginary axes represented space) (Hamilton, 1853, 1967: p. 152).[5]

Now, if Hamilton rejected a purely philological and a purely instrumentalist view of algebra and analysis and insisted on developing the latter as a *science* of pure time in the Kantian sense, i.e. presumably as based on synthetic *a priori* truths derived from pure intuition of time, is it likely that he would have an instrumentalist view of physical theories?

In so far as Kant's philosophy affirms that science is concerned with the study of the phenomenal world and is unable to provide us with knowledge of the "nature of things" or of "things in themselves", whether it is matter, light, or forces, etc., it favours a view of scientific theories which is close to Ramsey's (and Duhem's). Close but not identical, for there is another element in Kant's epistemology which separates it sharply from phenomenalism, viz. its emphasis on the active role of our mind in acquiring knowledge. Experience is possible only in terms of space and time as *a priori* forms of sensibility (or forms of outer and inner intuition respectively) which our mind imposes on what is given through the senses. Moreover, perception is not identical with knowledge. The latter consists of judgements—primarily of the so-called synthetic *a priori* kind—made by our faculty of understanding. Judgements involve concepts in terms of which objects of perception are constituted as thought objects. Thus *a priori* principles are "constitutive" of the objects with which our knowledge is concerned. Kant's doctrine of the active role of the mind and of two sources and components of knowledge proved very attractive to many creative mathematicians and physicists during the nineteenth century. Some—for example Hamilton—took it in the original, orthodox version; others, for example, Kirchhoff, Hertz, Poincaré modified it considerably. Common to all of them was the belief that knowledge is not received passively through our senses but results through an active operation of our mind;[6] science—physics in particular—is an organic fusion of mathematics and experimentation and the content of a physical theory is expressed not only in the observational consequences (framed in mathematical language of metric concepts) but also through the form of mathematical equations used to formulate the theory; both exclude certain "possible worlds" or models (to use a modern term). Unlike the instrumentalist doctrine attributed to Duhem and Ramsey, the view which we now want to attribute to Hamilton (as a Kantian) and even more emphatically to Poincaré and Hertz (Hertz, 1894, 1956: pp. 175-7), does not imply—in modern terminology—that theoretical terms are uninterpreted symbols (or merely formal instruments). Rather it stresses that the cognitive role of the abstract (theoretical) part of a theory consists in postulating a structure which reflects the structure of the domain(s) of application of the theory, or—in another terminology—which imposes relational restrictions on admissible interpretations of theoretical terms and thus reduces the class of possible models of the theory. Two rival theories are distinguishable one from another only up to the similarity of their formal structure, unless they also yield different observational predictions. This applies, however, not only to rival theories; similarity of structure is often the basis for the unification of apparently distinct theories (cf. Maxwell's electromagnetic theory). The epistemological implications of this doctrine, which may be labelled *conventionalist structuralism* or *conventionalism* for short, were elaborated by Poincaré and amount to the claim that reality is knowable only up to the observational equivalence and up to the isomorphism of the abstract postulates of apparently rival theories (Giedymin, 1978). Alternatively, in ontological (semantical) terminology: any two *similar worlds*, i.e. such that one can pass from one to another either by changing the coordinate axes, or by changing the scale of lengths, or by any point transformation, are *indistinguishable* (Poincaré, 1905: p. 39).

The following mathematical concepts proved especially important for the evolution of conventionalist structuralism between Hamilton's and Poincaré's time. The concept of a *higher space* is implicit in Hamilton's optics (8-space). Later on he used a 4-space to define the multiplication of quaternions but claimed "no originality as respects the paradox of the fourth dimension of space" and conceded priority to Cayley in constructing a geometry of four dimensions (Hamilton, 1967: p. 108; letter to H. Lloyd, 1844). It was J. Plücker who pointed out that the dimensionality of space depends on our *choice* of the elements (points, lines, planes, etc.) from which space is construed. He also pointed out that a right line may be construed in two different ways and that, accordingly, there are two constructions of space: in one, space is traversed by lines themselves consisting of points, in the other it is traversed by lines determined by planes passing through them. The first is used in optics when luminous points are assumed to send rays in all directions, the second when instead of rays one considers wave-fronts and their consecutive intersections (Plücker, 1865: pp. 725–6). The choice of the space element in this sense and hence of the dimensionality of space is discussed by Poincaré in (1898: pp. 21–4):

> ... Suppose that we choose the different transformations of a rotative sub-group. We have here a triple infinity. The material of our group is accordingly composed of a triple infinity of elements. The degree of the group is three. We have then chosen the point as the element of space and given to space three dimensions. Suppose we choose the different transformations of a helicoidal sub-group. Here we have a quadruple infinity. The material of our group is composed of a quadruple infinity of elements. Its degree is four. We then have chosen the straight line as the element of space—which would give to space four dimensions. Suppose, finally, that we choose the different transformations of a rotative sheaf. The degree would then be five. We have chosen as the element of space the figure formed by a straight line and a point on that straight line. Space would have five dimensions. Here are three solutions, which are each logically possible. We prefer the first because it is the simplest, and it is the simplest because it is that which gives to space the smallest number of dimensions. But there is another

reason which recommends this choice. The rotative sub-group first attracts our attention because it conserves certain sensations. The helicoidal sub-group is known to us only later and more indirectly. The rotative sheaf, on the other hand, is itself merely a sub-group of the rotative sub-group.

Essential to Hamilton's optics and wave dynamics were *canonical* or (infinitesimal) *contact-transformations* as mathematical representations of advancing wave-fronts (Whittaker, 1904: pp. 292–3, 1937: p. 292). One property of these transformations when interpreted geometrically is that they transform tangent surfaces into tangent surfaces and preserve the order of contact (hence the name invented by Sophus Lie); another that they may establish correspondence between different categories of space elements, e.g. they may transform a point (and directions issuing from it) into a surface and all the normals to the surface; or a line into a surface or into a point. The line-sphere transformation of Lie illustrates this interchange of categories as does duality in projective geometry. Lie's theory of groups of (continuous) transformations exerted considerable influence on Poincaré's view of geometry and physics (Giedymin, 1977). Both geometry and physics were seen by Poincaré as studies of the invariants with respect to certain transformation groups. Physics (and celestial dynamics) was treated as quasi-geometry in a space of $2n$ dimensions.

Hamilton and Poincaré

The impact on Poincaré's mathematical research and philosophical ideas of Hamilton's main contributions to mathematics is relatively easy to establish. On the other hand, the question whether Poincaré was familiar in any detail with and took notice of Hamilton's philosophical doctrines is a matter of conjecture.

At the time of his death in 1865 Hamilton was known outside Britain mainly as the author of the theory of quaternions and as the author of a new method of analytical dynamics, viz. the representation of the integral equations of motion which involve generalised coordinates and generalised momenta. His writings on geometrical optics, published in the *Proceedings of the Royal Irish Society*, were not easily accessible in Continental Europe and appear largely unknown (Klein, 1926, 1956: pp. 197-8) although the main ideas which they contain were reported by Hamilton in his papers on dynamics (Hamilton, 1834a, b, 1940). One of his characteristic functions, *T*, was rediscovered by Bruns and is often referred to as "angle eikonal".

The calculus of quaternions was an extension of Hamilton's earlier theory of algebraic couples (Hamilton, 1837, 1967: pp. 3-97) and both were aimed at providing a rigorous foundation for the theory of negative and imaginary numbers. Hamilton, according to his own account (Hamilton, 1853, 1967: p. 135), was inspired by the "geometrical representation of the square roots of negative quantities" given by John Warren (1828) and George Peacock, a representation which is usually attributed to Argand (Argand diagram). The idea was to extend to three-dimensional space the role that complex numbers play on a plane.

On the plane the complex number $x + iy$ represents both the point x, y and the directed line segment from 0 to the point x, y, i.e. a vector. The addition of two complex numbers

$$(x + iy) + (a + ib) = (x + a) + i(y + b)$$

is represented as the addition of vectors, i.e. as a parallel translation of the plane expressed by $a + ib$. The multiplication

$$(x + iy) \cdot (a + ib) = (x + iy) \cdot \rho \cdot e^{i\varphi}$$

effects a rotation of the plane about the point 0 by the angle φ with simultaneous stretching of the segments in the proportion $1:\rho$.

In his attempt to extend the operations on directed lines from the plane to the three-dimensional space Hamilton noticed that, although addition could be treated analogously, multiplication constituted a problem the solution of which took him a long time and involved the rejection of the commutative law of ordinary algebra. In several accounts of his discovery, Hamilton (1833c, 1853) emphasised very strongly the role played in it by analogy and by general principles, especially the principle of the symmetry of space ("no one direction of space is to be regarded as eminent above another").

From one point of view the calculus of quaternions is a four-element non-commutative algebra, the elements l, i, j, k satisfying the equations:

$$i^2 = j^2 = k^2 = -1; \quad ij = -ji = k; \quad jk = -kj = i; \quad ki = -ik = j$$

and the general quaternion being the quadrinomial $w + ix + jy + kz$ (Hamilton, 1853, 1967: pp. 142–3). Although Hamilton himself did not use group-theoretic terminology, we can say that the elements $\pm 1, \pm i, \pm j, \pm k$, satisfying the above equations form a non-commutative group (of order 8) under multiplication, which is now called the quaternion group.[7]

From the geometrical viewpoint the theory of quaternions is a calculus of vectors in 3-space. Vectors are treated in it directly as space elements without the intervention of Cartesian coordinates[8] and a quaternion is a four-parameter operation that changes one vector into another: the tensor of a quaternion (not to be confused with the tensor of the tensor calculus) is the stretching or shrinking parameter (and is a number, i.e. a scalar); the versor of a quaternion is a three-parameter operation of turning one unit vector into another and may, therefore, be regarded as a rotation of space about a fixed point. The terms scalar, vector and tensor (of a quaternion) were introduced for the first time by Hamilton. In his (1845) Cayley acknowledged Hamilton's discovery of the formula for three-dimensional rotation in terms of quaternions and showed that quaternions can be used to represent rotations in a four-dimensional space.[9] In Cayley's theory of matrices (1858) quaternions occur as a special case; they are isomorphic to 2×2 matrices with complex elements. Quaternions turned out to be intimately linked with binary linear substitutions.

Ideas similar to those of Hamilton's quaternion calculus had been used earlier by Gauss and occurred almost simultaneously in a work by Olinde Rodrigues (1840) and by Grassmann (1844).[10] Specific to Hamilton's work is the introduction of the concept of *field*, of great importance for the applications of quaternions in physics. With each point in space a quaternion is associated, i.e. a scalar and a vector. From such a quaternion field $t(x, y, z) + iu(x, y, z) + jv(x, y, z) +$

$kw(x, y, z)$ further fields are obtained by suitable operations. The "Nabla" operator denoted by ∇ and defined as

$$\nabla = i\frac{\delta}{\delta x} + j\frac{\delta}{\delta y} + k\frac{\delta}{\delta z}$$

is used to introduce all the fundamental concepts of vector analysis, such as $\nabla t = \text{grad } t$ (where t is a scalar), the *divergence* (Div) of the field and *curl*, the former being the scalar and the latter the vector of $\nabla(iu + jv + kw)$; when applied twice to a scalar the operator ∇ gives $-\Delta$ (delta):

$$\nabla^2 t = -\Delta t = -\frac{\delta^2 t}{\delta x^2} + \frac{\delta^2 t}{\delta y^2} + \frac{\delta^2 t}{\delta z^2}.$$

The almost religious fervour with which Hamilton in his later years and some of his followers pursued the research on quaternions provoked not only criticism of the more excessive claims concerning the importance and fruitfulness of the theory but also opposition and even ridicule. According to Whittaker (1940: p. 154):

> ... When Hamilton died in 1865, at the age of sixty, his rank in the hierarchy of the great men of science was very uncertain. His name was at that time associated chiefly with the discovery of quaternions: and whether quaternions had much value was a subject of discussion, and often of acrimonious controversy, for the next fifty years, the general trend of opinion becoming on the whole less favourable to them. The nadir of Hamilton's reputation was touched about the beginning of the present century: since when, there has been a steady movement in the reverse direction: one after another, the significance of his great innovations has been appreciated: and to-day he is placed among the greatest, in the class inferior only to Newton.

Against the background of the above statement it is interesting to read Poincaré's appreciation of quaternions and of vector analysis in the development of pure and applied mathematics.

In a brief historical introduction to his review of Hilbert's *Grundlagen der Geometrie* in (1902a, 1956: p. 93) Poincaré wrote:

> ... The operations of arithmetic in turn had been subjected to criticism and the quaternions of Hamilton have shown us an example of an operation which

presents such a perfect analogy with multiplication that one can give it the same name and which, nevertheless, is not commutative, i.e. whose result is not independent of the order of the factors. This was, in arithmetic, a revolution quite similar to the one brought about by Lobatchevsky in geometry.

In "Analysis and physics" (1905) Poincaré was concerned with the "mutual interpenetration" of pure analysis and mathematical physics. Having made the point that pure analysis practised as an art furnishes the physicist with "the only language he can speak" (1905, 1958: pp. 76–7), Poincaré went on to say that the interdependence of the two disciplines is much deeper. For the language of analysis transforms an individual experiment into general laws. This can be done, however, in an infinite number of ways. We are guided in this choice by *analogy*. But what kind of analogy? The analogy that is relevant here—so Poincaré continues (1905, 1958: p. 77)—is mathematical or formal:

> ...Primitive man knew only crude analogies, those which strike the senses, those of colours or of sounds. He never would have dreamt of likening light to radiant heat. What has taught us to know the true, profound analogies, those the eyes do not see but reason divines? It is the mathematical spirit, which disdains matter to cling only to pure form. This it is which has taught us to give the same name to things differing only in material, to call by the same name, for instance, the multiplications of quaternions and that of whole numbers. If the quaternions, of which I have just spoken, had not been so promptly utilised by the English physicists, many persons would doubtless see in them only a useless fancy, and yet, in teaching us to liken what appearances separate, they would have already rendered us more apt to penetrate the secrets of nature.[11]

As an example of how analysis helps us "to see, to discern our way in the labyrinth", Poincaré mentions Maxwell's contributions to electro-magnetism. It was not a new experiment which Maxwell used to modify the existing laws of electro-dynamics. Rather by looking at them "under a new bias" Maxwell saw that the equations became more symmetrical when a term was added, viz.

$$\frac{1}{c}\frac{dE}{dt}$$

(see also Chap. 5, this volume), a term which in any case was too small to produce effects appreciable with the help of old experimental methods so that one had to wait twenty years for experimental verification. How was this triumph obtained, Poincaré asks, and gives the following reply:

> ... It was because Maxwell was profoundly steeped in the sense of mathematical symmetry; would he have been so, if others before him had not studied this symmetry for its own beauty? It was because Maxwell was accustomed to "think in vectors" and yet it was through the theory of imaginaries (neomonics) that vectors were introduced into analysis. And those who invented imaginaries hardly suspected the advantage which would be obtained from them for the study of the real world; of this the name given them is proof sufficient (1905, 1958: p. 78).

Two closely interrelated aspects of Poincaré's appreciation of Hamilton's theory of quaternions should be emphasised here in order to bring out common features in the works of the two mathematicians and in order to focus on those contributions of Hamilton which influenced Poincaré's thinking in mathematics and in philosophy. The first of these aspects concerns the creation of quaternions as a kind of paradigm of mathematical discovery in which—to use Poincaré's philosophical terminology—the form of diverse mathematical objects is "divined by the mind". The second is concerned with the ability of quaternions and of the language of vector analysis to express symmetry (invariance) in physics.

Naturally Poincaré saw Hamilton's quaternions first of all as the discovery of a non-commutative algebra and hence as a turning point in the development of algebra. At the same time, however, he saw in it a paradigm of mathematical discovery in which apparently different objects (numbers, sets of numbers, operations on either) are identified on the basis of their *form* (formal features, formal analogy "divined by the mind") so that they can be given the same name and subsumed under a general theory, in this case linear associative algebra over the field of real numbers. Poincaré's own first major mathematical discovery, viz. the discovery of automorphic functions, was also of this type (Poincaré, 1881, and eleven memoirs between 1882 and 1898; see bibliography in Ford, 1951). "An automorphic function" is a generalisation of "a periodic function". A circular function, e.g. *sinz*, has the property that it remains unchanged in value if z is subjected to a transformation of the group $z' = z + 2m\pi$; a hyperbolic function, e.g. $\sinh z$ is unchanged in value if subjected to a transformation of the group $z' = z + 2m\pi i$; an elliptic function, e.g. Weierstrassian function $\wp(z)$ remains unchanged in value under transformations of a group of the form $z' = z + m\omega + m'\omega'$. Poincaré discovered that periodicity is a special case of a more general property: certain functions $f(z)$ are such that they remain unchanged in value when subjected to any of a (denumerably infinite) group of fractional transformations of the form

$$z' = \frac{az + b}{dz + d}.$$

A transformation of this form where a, b, c, d, are constants and $ad - bc \neq 0$ is a linear fractional transformation. A function $f(z)$ whose value is unchanged under transformations of this group is called *automorphic*. Poincaré not only established the foundations of the theory of automorphic functions but also discovered its connection with hyperbolic geometry. His own account of this discovery (Poincaré, 1908, last Chap.) is not unlike Hamilton's account of the discovery of the multiplication rule for quaternions. Poincaré reports that during a geological trip sponsored by the School of Mines in which he took part "... reaching Coutances we took a bus for some excursion or another. The instant I put my foot on the step the idea came to me ... that the transformations which I had used to define Fuchsian functions [a class of automorphic function, J.G.] were identical with those of non-Euclidean geometry...." At any rate, whatever the psychological circumstances of the discovery, the connection between automorphic functions and hyperbolic geometry is this: If we consider Poincaré's model of plane hyperbolic geometry (known already to Beltrami) and take the interior of the unit circle as the domain of an automorphic function, then the associated linear substitutions transform the unit circle into itself and can be regarded as movements of the hyperbolic plane.[12] More generally, one can also show that the group of hyperbolic movements of space is isomorphic with the group of (proper) linear substitutions of a complex variable (Klein, 1927: p. 308). Finally, Poincaré (and Klein) established connections between groups of linear fractional transformations and the solutions of linear homogeneous differential equations.[13] In this way two abstract mathematical theories, the theory of automorphic functions and hyperbolic geometry, became connected with a branch of mathematics which has direct applications in physics.

The symmetry of physical laws formulated in the language of vectors was one of the main concerns of Poincaré in his two papers on relativity theory (Poincaré, 1905a, 1906) in which—among other things—he proved the covariance of the Maxwell–Lorentz equations under (special) Lorentz transformation, established that Lorentz transformations form a group, investigated the invariants of the group and showed that Lorentz transformation may be interpreted geometrically as rotation in four-dimensional space (space-time) (Poincaré, 1905a, 1906, 1970: pp. 150–85; Kilmister, 1970, Introduction). In doing this Poincaré used so-called four-vectors whose introduction is usually attributed to a later paper by H. Minkowski (1907, 1915) (Lorentz, 1915, 1921, 1954: pp. 683–701).

Hamilton's method in dynamics affected not only Poincaré's research in mathematical physics and astronomy (for example, Poincaré, 1890, 1896, 1897, 1901, 1909) but also his views on the structure and epistemological status of physical theories as well as his views on the rationality of scientific change, in short his (conventionalist) theory of scientific knowledge.

The discovery of Hamilton's method of geometrical optics was due, as we know, to formal similarities between the (Maupertuis) principle of least action and the (Fermat) principle of least time. Within the emission theory of light, the Maupertuis Principle subsumed optics under Newtonian dynamics. Hamilton conceived the daring idea of generalising his method in optics to dynamics and of constructing dynamics "as a kind of optics", to put it in terms used later by some authors (for example, Klein, 1926, 1956: p. 198; also Synge, 1937). Since certain functions in Hamilton's formulation of the theory generate contact (or canonical) transformations which are also employed to represent the propagation of wave fronts, in effect the formalism of wave optics was thus extended to dynamics.

As the title of his (1834a) indicates, Hamilton's method in dynamics consists in "the search and differentiation of one central relation, or characteristic function" of a dynamic system. In his first two memoirs (1834, 1835) Hamilton introduced in fact three such functions, viz. V, S and H in terms of the two constants of motion, T and U (the kinetic and potential energies) known from Lagrangean equations—and taken as functions of coordinates, their derivatives, and time. Hamilton's Principal Function S is given by the equation:

$$S = \int_0^t (T - U)\, dt \tag{1}$$

or, if we use the Lagrangean function $L = T - U$,

$$S = \int_0^t L\, dt \tag{1'}$$

where the integration is over the actual path of the system. The variational principle

$$\delta S = 0 \quad \text{or} \quad \delta \int_0^t L\, dt = 0 \tag{2}$$

called by Hamilton the Principle of Stationary Action now bears his name. It was implicit in Lagrange's theory but never stated in it explicitly. One of the features which distinguish Hamilton's version of analytic dynamics from the Lagrangean is the use of generalised momenta along with Lagrangean generalised coordinates in the description of the state of a dynamical system. Although these two quantities seem conceptually quite different, within Hamilton's theory the variables representing them are given the same status, i.e. are treated symmetrically. This makes the theory itself more abstract, more general and flexible. If we use q_i for generalised coordinates, p_i for generalised momenta and $\dot{q}_i$, $\dot{p}_i$ for their time derivatives—the symbolism introduced in subsequent nineteenth-century developments of analytical dynamics (cf. Cayley, 1858: p.

14)—then generalised momentum components are defined by:

$$p_i = \frac{\delta L}{\delta \dot{q}_i} \text{ so that } \dot{p}_i = \frac{\delta L}{\delta q_i} \ (i = 1, 2, \ldots, n). \tag{3}$$

Introducing for the total energy $T + U$ the expression $H(p_i, q_i, t)$ defined by

$$H = \sum_i p_i \dot{q}_i - L \tag{4}$$

we can express Hamilton's differential equations of a dynamical system in the following simple so-called *canonical*, form:

$$\dot{q}_i = \frac{\delta H}{\delta p_i}, \quad \dot{p}_i = -\frac{\delta H}{\delta q_i}. \tag{5}$$

If in Hamilton's variational principle we substitute for L the right-hand side of the equation:

$$L = \sum_i p_i \dot{q}_i - H \tag{6}$$

then Hamilton's differential equations (canonical equations) are derivable from the Hamiltonian principle thus modified.

Lagrange's equations are invariant with respect to point transformations which have the form $q'_i = q'_i(q, t)$. Since in Hamilton's equations, apart from generalised coordinates q_i also generalised momenta p_i occur, a more general concept of transformation must be used. The transformation equations must have the form:

$$\begin{aligned} q'_i &= q'_i(q, p, t), \\ p'_i &= p'_i(q, p, t). \end{aligned} \tag{7}$$

Carl Jacobi (born 1804) was the first to raise the question of finding the most general substitutions of this sort which do not change the form of canonical (Hamilton's) equations. His answer was that the transformed variables p'_i, q'_i are related to the original p_i, q_i by such canonical transformations (or contact transformations) if and only if there exists a (differentiable) function Ω of q_i and q'_i, called the generating function of the (canonical) transformation, and such that the following relations hold:

$$\frac{\delta\Omega}{\delta q_i} = p_i \quad \frac{\delta\Omega}{\delta q'_i} = -p'_i. \tag{8}$$

Canonical transformations form a group, hence we may say that Hamilton's differential equations of dynamics are form-invariant (or are covariant) with respect to the group of canonical transformations. The same applies to Hamilton's modified Principle. Among the invariants of the group of canonical transformations are Poincaré's integral invariants and Lagrange and Poisson bracket-expressions (Whittaker, 1904, 1937: pp. 267–76, 298–302).

This short outline of Hamilton's contributions to analytical dynamics is relevant to three doctrines which are part of Poincaré's conventionalist epistemology: the conception of mathematical physics and astronomy as (quasi-)geometry, the conception of the cognitive content of a theory as consisting of observational consequences and of its mathematical form or structure and the idea of progress in science as a sequence of content-preserving theoretical changes.

Physical laws are usually expressed as differential equations and differential equations and their integration may be given geometrical interpretation. Indeed, to inquire into the geometrical sense of differential equations and their integration is a task of differential geometry as conceived already by Gaspard Monge (1746–1818) and subsequently by Sophus Lie. So, for example, a partial differential equation of the first order in x, y, z, is equivalent to the problem: "To find the most general surface which at every one of its points touches a cone associated with that point, the general equation of the cone in plane coordinates being represented by the given partial differential equation" (Lie, 1871, 1958: pp. 491–2; see also Klein, 1926, 1957: pp. 71–4). In his theory of differential equations Lie established the relation between differential equations, groups which they admit and the integral curves defined by the solution functions of the equations. The group with respect to which an equation is form-invariant (the group which the equation admits) is the group with respect to which the family of the integral curves of the equation is also invariant.

In his (1858) "Report on the recent progress of theoretical dynamics", Arthur Cayley refers to the contributors to the development of analytical dynamics between 1788 and 1857 as geometers and in his Presidential address (1884) he applies the term "quasi-geometry" to non-Euclidean geometry and to the geometries of higher spaces. In Lagrange's theory the motion of a dynamical system of n-degrees of freedom is represented by the trajectory of a point in 3-n "configuration space". In Hamilton's theory the introduction of generalised momenta along with generalised coordinates results in the use of 6-n "phase space" for the same purpose. Canonical transformations transform the phase space into itself. Analytical dynamics is seen from this point of view as the study of the invariants with respect to the group of canonical transformations. This is analogous to the Lie–Klein–Poincaré view of geometry as the study of continuous (topological) groups and is the basis of Poincaré's extension of geometrical conventionalism to physics. An obvious conventionalist feature in this approach to dynamics, strongly emphasised by Poincaré for example in (1898: p. 24) but ultimately due to Plücker, is the choice of the dimensionality of space. A description of the motion of a dynamical system in the 6-n dimensional phase space is readily translatable into the language of ordinary three-dimensional Euclidean space with Cartesian coordinates. Moreover, though the Lagrangean configuration space and the Hamiltonian phase space are of different dimensions and though Lagrangean equations of motion are second-order differential equations whereas Hamilton's equations are first-order differential equations, the two theories are observationally equivalent wherever both apply. Finally, the transformation theory of dynamics has an important pragmatic role since it is essential to the procedure of solving dynamical problems. The integration of a system of equations soluble by quadratures (i.e. in terms of known elementary functions or of their indefinite integrals) may be affected by transforming it into another dynamical system with fewer degrees of freedom. This can be done in some cases by the procedure known as the ignoration of coordinates, which is based on a specialisation of a general theorem to the effect that any integral

$$\phi(q_1, q_2, \ldots, q_n, p_1, p_2, \ldots, p_n, t) = \text{Constant}$$

corresponds to an infinitesimal contact transformation whose symbol is the Poisson-bracket (ϕ, f); or to put it informally: integrals of a dynamical system and contact transformations which change the system into itself (or: are admitted by the system) are substantially the same thing (Whittaker, 1904, 1937: pp. 319-20).

The historical evolution of the interest in geometrical transformations is important to the understanding of Poincaré's conventionalist epistemology. At first transformations were seen mainly as devices which simplify problems or make otherwise intractable problems solvable. This is very clear in the philosophy underlying Poncelet's construction of projective geometry (Poncelet, 1822). The procedure which he recommends for finding properties of classes of figures from simpler special cases is this. Having ascertained projectively invariant properties one should, firstly, simplify the figure by central projection, secondly, prove a theorem for the simplified figure, thirdly, formulate the result in projectively invariant form (cf. Coolidge, 1963: p. 268). Similar ideas of a pragmatic nature may be found in Plücker's philosophy of geometrical transformations in which so-called *Übertragungsprinzipien* take advantage of the ambiguity of symbols allowing interpretations in terms of different types of coordinates so that one and the same equation represents different geometrical figures (Plücker, 1835). Plücker's clarification of the principle of duality in projective geometry was in the same spirit as Cayley's discussion of the duality principle (Cayley, 1859). In time the attention of mathematicians shifted from what may be achieved with the help of transformations to the transformations themselves, their group properties and invariants, the variety of groups, etc.—in other words, to what Poincaré calls in his (1889) the "form" of a group in contradistinction to its "material". In this way groups become the focus of interest as one of the fundamental types of mathematical structure and group theory becomes a unifying ordering and clarifying theory which provides a conceptual methodological and even philosophical foundation for the study of mathematical structures in pure and applied mathematics. In Klein's *Erlanger Programm* (1872) the mathematical side is stressed almost exclusively. Epistemological considerations regarding the foundational role of transformation groups were introduced mainly by Poincaré whose initial inspiration in that area came in any case from Sophus Lie rather than from Klein. The pragmatic role of transformations in problem solving as well as the idea of invariants with respect to transformation groups as descriptive of structural features of collections of objects, became essential ingredients of Poincaré's conventionalist epistemology of geometry and of physics as quasi-geometry. (Pragmatist elements in Poincaré's conventionalism were strongly emphasised by Berthelot (1911).) Hamilton's theory of dynamics was particularly important here because contact (canonical) transformations are the basis of the integration of Hamilton's canonical equations and of the form of Hamilton's equations themselves. Since contact transformations may be either one-sorted or many-sorted, i.e. they establish correspondence between either objects of one sort or of different sorts, this seems to have encouraged Poincaré's view that conceptual and ontological changes in mathematics and in science are not as disruptive and as inimical to continuity and objectivity in scientific knowledge as may appear at first sight.

Another doctrine which became part of Poincaré's conventionalist epistemology and which reflected some of the main developments in nineteenth-century mathematical physics, concerned the structure and the cognitive content of a physical theory. The theories of analytical dynamics of Lagrange, Poisson, Hamilton, Jacobi and others, Fourier's theory of heat, Hamilton's and Cauchy's optics, as well as Maxwell's theory of electromagnetism, constitute paradigms of a type of physical theory to which in 1904 (1905) Poincaré gave the name of "physics of the principles" in order to contrast it with another style of theorising, typified by Laplace's celestial mechanics—or the "physics of central forces" in general.[14] Whereas physicists in the second category, especially Laplace and his followers, believe that the aim of physics is to "penetrate the mystery of the universe" and to provide deterministic theories which not only predict observable effects but also postulate some hidden mechanism behind the phenomena, physicists in the first category engage in the construction of mathematical theories based on very general assumptions such as the principle of conservation of energy, of entropy, of least action, Hamilton's principle, etc., in other words on assumptions which yield desired observable predictions from suitable initial conditions *without* making any explicit reference to hidden mechanisms and *yet* are reconcilable with many—often mutually incompatible—theoretical explanations. The distinction between these two types of theory (and theorists) had in fact been made by Poincaré at an earlier time, viz. in his 1888–9 lectures on electricity and optics (Poincaré, 1890, Introduction) one of the main objects of which was to introduce and explain to French students Maxwell's electromagnetic theory. According to Poincaré's account, having failed to provide a satisfactory mechanical model of electromagnetic phenomena Maxwell in his main treatise had to be satisfied with a different procedure. He showed, namely, that a mechanical explanation of a set of facts is possible if and only if there exist two functions, T and U, such that they satisfy the principle of conservation of energy (i.e. for an isolated system $T + U =$ constant) and the principle of least action (Poincaré, 1890: Chaps. XII, VI–VII; 1902, 1952: p. 220; 1905, 1958: p. 93).[15] However, by a theorem due to Königs one can show that if there is one such explanation, there are indefinitely many—either postulating discrete unobservable objects or an unobservable, continuous medium such as the ether (Poincaré, 1897: pp. 248–9; 1890: pp. xiv–xv; 1902, 1952: pp. 167–8). The choice of one of these explanations cannot—under these circumstances—be made on experimental grounds and Maxwell rightly abstained from making a choice. Hence Maxwell's theories are just Maxwell's equations. Poincaré concludes with this comment about Maxwell's method:

> ...He throws into relief the essential, i.e. what is common to all theories; everything that suits only a particular theory is passed over almost in silence. The reader therefore finds himself in the presence of form nearly devoid of matter, which at first he is tempted to take as a fugitive and unassailable phantom. But

the efforts he is thus compelled to make force him to think, and eventually he sees that there is often something rather artificial in the theoretical "aggregates" which he once admired. (Poincaré, 1890: Chap. XVI; 1902, 1952: p. 224).

Is the "physics of the principles" phenomenalist? It *was* seen as such by the adherents of energeticism, Rankine, Helm, Ostwald, Duhem and others who were opposed to mechanistic philosophy and tried to eliminate from science "atoms", "ether" and determinism as metaphysical. However, this was not the view of those mathematicians like Lagrange, Poisson, Fourier, Hamilton, Jacobi who were engaged in constructing the physics of the principles. Nor was it the view held by Poincaré,[16] To be sure, what they aimed at and put into effect *was* a replacement programme. However, unlike energeticists, they did *not* desire to replace mechanistic theories by low-level empirical generalisations but rather by theories which make free use of sophisticated mathematics and have a wide—in fact, unlimited—range of applications. Moreover, their replacement programme—unlike that of energeticists—was *not* motivated by the wish to eradicate theoretical concepts and postulates as "meaningless" in the light of some extreme empiricist or positivist criterion of meaningfulness. Rather, their aim was to provide theories which—while observationally adequate (as far as one could see at the time)—would be *elevated*, by the use of abstract or ambiguous mathematical symbols, *above* current theoretical or ontological disputes so that one could use and further elaborate them no matter which side in the dispute (if any) one favoured or which side (if any) would eventually win the day. In other words, the aim was to provide theories "which remain true whatever may be the details of the invisible mechanism..." (Poincaré, 1905, 1958: p. 94). There are several reasons why this replacement programme appeared desirable. One of the reasons is convenience and economy of thought. So, for example, it is convenient for certain purposes to be able to apply Hamilton's optics to an optical instrument and ignore the latter's complicated internal structure. Another reason is continuity of research even when the opposition between rival theories stubbornly resists experimental resolution over longer periods of time. This was the case, for example, with rival theories of light. There is also

Maxwell's reason, which frequently occurs in research (the new quantum theory may be an example), viz. failure, in spite of persistent efforts, to find a theoretical explanation which satisfies certain requirements fixed in advance.[17] At any rate, it should be obvious that the physics of the principles, in spite of its replacement programme has little in common with the instrumentalist view of theories as characterised, for example, by Nagel (1961, Chap. 6) or Popper (1963, Chap. 3). Essential to the latter seems to be either an ontology which identifies the class of existing objects with the class of observable objects (with macro-objects) or else a criterion of meaning which identifies meaningful expressions with observables. Neither is part of the philosophy underlying the physics of the principles. However, from what has been said so far it does seem plausible to regard the theories based on variational, conservation and invariance principles as having the Ramsey–sentence form. Moreover, there are passages in Poincaré's writings which appear to indicate that he saw generally the structure of theories in this way. So, for example, when discussing the transition from Fresnel's optics to Maxwell's electromagnetic theory Poincaré writes:

> "... (Fresnel's) object was not to know whether there really is an ether, if it is or is not formed of atoms, if these atoms really move in this way of that; his object was to predict optical phenomena. This Fresnel's theory enables us to do to-day as well as it did before Maxwell's time..." (Poincaré, 1902, 1952: pp. 160–1).

Passages, like the above, have been used by commentators to attribute to Poincaré an instrumentalist (or phenomenalist or pragmatist) view of theories. What seems to have been overlooked is the anticipation by Poincaré and repudiation of such suggestions. Poincaré continues the quoted passage by saying that the view expressed in it does *not* reduce physical theories to simple practical recipes, for an essential component of a theory is a set of equations and "equations express relations":

> ",,, They teach us now, as they did then, that there is such and such a relation between this thing and that; only the something which we then called *motion*, we now call *electric current*. But these are merely names of the images we substituted for real objects which Nature will hide for ever from our eyes..." (1902, 1952: p. 161).

In other words, according to Poincaré, the cognitive content of a physical theory is determined not only by its observational consequences but also by the relational structure expressed through its equations. The former represent the phenomena, the latter "A profound reality" (Poincaré, 1902, 1952: pp. 161–2). Both express "... the truth which will ever remain the same in whatever garb we may see fit to clothe it" (Poincaré, 1902, 1952: p. 162). There seems to be little difference between this view of theories and the view of realists like Ludwig Boltzmann who in his (1897, 1974: p. 226) argued against phenomenalism in this way:

> People have often put forward as ideal the mere setting up of partial differential equations and prediction of phenomena from them. However, they too are nothing more than rules for constructing alien mental pictures, namely of series of numbers. Partial differential equations require the construction of collections of numbers representing a manifold of several dimensions. If we remember the meaning of their symbolism they are nothing more than the demand to imagine very many points of such manifolds (that is, positions that are characterised by several numbers of the manifold, as spatial points are by their co-ordinates) and, using certain rules, constantly to derive from them new points of the manifold, to imagine, as it were, a progressive movement of the points in the manifold. Thus if we go to the bottom of it, Maxwell's electromagnetic equations in their Hertzian form likewise contain hypothetical features added to experience.... The assertion that atomism does, while partial differential equations do not, introduce material extraneous to the facts seems to me unfounded....

If in the quoted passage we ignore the dubious assertion that mental pictures are invariably associated with (or are even necessary for) the use of mathematical symbols, then the following claims remain: (a) physics is a kind of (quasi-)geometry; (b) in using differential equations we transcend the strict phenomenalist programme; (c) the mathematics of the abstract part of a physical theory contributes to its descriptive content; (d) hypothetical assumptions such as that of continuity are implicit in the use of differential equations (which involve time as an independent variable); (e) any differential equation as part of a physical theory is just as hypothetical as is atomism.

Poincaré clearly accepted (a), (b) and (c). On the other hand, in response to (d) and (e) he pointed out in 1900 in "Sur les rapports de la physique expérimentale et de la physique mathématique" (1902,

1952: pp. 152-3, 160-1, 162-9) that, firstly, the mathematical assumptions of continuity or discreteness often play the role of "indifferent" hypotheses, which may replace one another without affecting the result, i.e. the content of the theory; secondly, either of these assumptions—unlike atomism (as a specific theory)—is compatible with an indefinite number of apparently rival but observationally and formally indistinguishable hypotheses. In other words, according to Poincaré, the equations forming the abstract part of a physical theory contribute to the theory's content not only through its observational consequences,[18] but also through the form of the equations used or through the form of their solution functions. By the form of the equations he meant their type (ordinary or partial differential equations), order and the group which they admit, i.e. the group with respect to which they are form-invariant (covariant). If two theories are not only observationally equivalent but also use equations of the same form, then they are indistinguishable or strictly equivalent.[19] (This may be called the conventionalist principle of equivalence or the thesis of structuralist realism.) If on the other hand observationally equivalent theories are expressed in equations of different form, then this formal difference *may but need not be used to assert descriptive difference* (compare the distinction between assertive and neutral axiomatic systems in Chap. 4 in the present volume). Whether it is or not depends on whether one adopts a realist (assertive) or a nominalist (neutral) attitude. So, for example, the assumption of continuity implicit in the mathematical formalism may but need not be asserted descriptively. Similarly, the difference in type and order of equations may but need not have asserted, descriptive significance. This is so, of course, provided that in our understanding of physical theories we take into account not only their formal (syntactical) and model-theoretic (semantical) properties but also pragmatic relations between theories and men (who invent or use them), as Poincaré did when he talked of "elevating an empirical law to the status of a conventional principle", or of "adopting a nominalist attitude", etc.

The idea of two theories which are observationally equivalent and have the same form also suggested to Poincaré a solution of the epistemological problem of objectivity which was of greatest concern to him. The problem, as Poincaré saw it, was the following. How is objective and continuous growth of scientific knowledge possible in spite of the "ephemeral nature of theories", i.e. in spite of frequent, apparently disruptive changes in mathematical and physical theories? In fact this was also the problem which guided those who constructed the physics of the principles, especially Lagrange, Fourier, Cauchy, Hamilton.[20] Poincaré's solution of this epistemological problem and, perhaps, the main positive thesis of his conventionalism, was this. Frequent changes in physics, which make the impression of discontinuity, in fact concern mainly those components of physical theories which, though useful and psychologically and pragmatically indispensable (scientific metaphors, "indifferent hypotheses" in the sense of Poincaré, 1902, Chap. IX) do *not* contribute to the cognitive content of a theory as previously defined in terms of the observational consequences and the formal structure of the theory or else are mutually replaceable without loss of content. The procedure to maintain the continuity of scientific progress, invented by the creators of the physics of the principles and absorbed into Poincaré's conventionalism as part of its descriptive and normative philosophy, may be characterised thus. When confronted with apparently rival theories which persistently defy experimental decision, find a more general theory and from this vantage point show that the rivals are either special cases, with restricted applicability, or else are intertranslatable on the basis of a suitable dictionary or of a mapping between their domains (models).[21] One example of this procedure is the projective approach to metric geometries which evolved from Poncelet's work through the contributions of Cayley and Klein, with Darboux transformations as the basis for the correspondence between the Euclidean and Lobatchevskian spaces (Darboux, 1864, 1873, 1894, 1914).[22] Similarly, the affine group and its invariants are the common basis for the Newton–Galileo group of the classical and the Lorentz–group of relativistic mechanics (Klein, 1910: p. 281). Another example may be seen in Hamilton's method in geometrical optics, afterwards generalised by him to dynamics.

Notes

* Pp. 71–106 appeared in *Prospects for Pragmatism* (edited by D. Hugh Mellor), Cambridge University Press, 1980, under the title "Hamilton's method in geometrical optics and Ramsey's view of theories".
1. Original dates of publication of Hamilton's papers are given. Page references are to *The Mathematical Papers of Sir William Rowan Hamilton*, Vol. I, 1931, Vol. II, 1940, Vol. III, 1967.
2. It is possible to treat geometrical optics as a limiting case of physical optics (the wavelength of light tending to zero). Hamilton, however, developed geometrical optics as an independent discipline.
3. Hamilton's original symbolism is retained here. It must be noted that he used the symbol δ also for partial differential coefficients.
4. See also Conway and Synge, Note 20, "On group velocity and wave mechanics", pp. 500–2 of Hamilton (1931). On the genesis of Schrödinger's wave mechanics from the ideas of Hamilton's optics, see *Annln. Phys.* 4, 79, 1926, pp. 489 et seq. On the effect of Whittaker's (1904) on the research in quantum theory, see McCrea (1957).
5. The following passages from the Preface to *Lectures on Quaternions* (1853, 1967) are relevant here:

    "[60] In this way, then, or in one not essentially different, the fundamental formulae [48] of the calculus of quaternions, as first exhibited to the R.I.A. in 1843, namely, the equations,

    $$i^2 = -1, j^2 = -1, k^2 = -1 \quad \text{(A)}$$

    $$ij = +k, jk = +i, ki = +j \quad \text{(B)}$$

    $$ji = -k, kj = -i, ik = -j \quad \text{(C)}$$

were shown (in 1844) to be consistent with *a priori* principles, and with considerations of a general nature; a *product* being *here* regarded as a *fourth proportional*, to a certain *extra-spatial*[1] unit, and to *two directed factor-lines* in space: whereas, in the investigation of paragraphs [50] to [56], it was viewed rather as a certain FUNCTION of those two factors, the *form* of which function was to be determined in the manner most consistent with some general and guiding analogies, and with the conception of the *symmetry of space*" [italics, etc, as in the original—J.G.].

Hamilton's footnote (1) to "extra-spatial" reads as follows:

"It seemed (and still seems) to me natural to connect this *extra-spatial unit* with the conception [3] of TIME, regarded here merely as an *axis of continuous and uni-dimensional progression*. But whether we thus *consider jointly time and space*, or consider generally *any system of independent axes*, or scales of progression (u, i, j, k), I am disposed to infer from the above investigation of the following LAW OF THE FOUR SCALES, as one which is at least consistent with analogy, and admissible as a *definitional extension* of the fundamental equations of quaternions: 'A formula of proportion between four independent and directed units is to be considered as remaining true, when any two of them change places with each other (in the formula), provided that the direction (or sign) of one be reversed'. Whatever may be thought of these abstract and semi-metaphysical views, the

formulae (A), (B), (C) of para. [60] are in any event a sufficient basis for the erection of a CALCULUS of quaternions" (Hamilton, 1853, 1967: p. 152).
6. Poincaré wrote (1898: p. 3):
"Such a classification [of sensations—J.G.] cannot be accomplished without the active intervention of the mind, and it is the object of this intervention to refer our sensations to a sort of rubric or category pre-existing in us. Is this category to be regarded as a "form of our sensibility"? No, not in the sense that our sensations, individually considered, could not exist without it. It becomes necessary to us only for comparing our sensations, for reasoning upon our sensations. It is therefore rather a form of our understanding".

The claim that all knowledge of which our understanding is capable is relational in nature was also expressed clearly by René Descartes in *Regulae ad Directionem Igenii* (first Latin edition, Amsterdam, 1701). In Rule X we read that all "human sagacity" consists in proper observance of systems "of order" and, in Rule XIV that "all knowledge is comparison". See also note 12.

7. It was proved later that the only linear associative algebras over the field of real numbers in which division is uniquely possible, are the field of real numbers, the field of ordinary complex numbers and real quaternions.

8. Hamilton reported about himself (1847: 1967: p. 441):
"... The author stated that during a visit which he had lately made to England, Sir John Herschel suggested to him that the internal character (if it may be so called) of the method of quaternions, or of vectors, as applied to algebraic geometry, that character by which it is independent of any foreign and arbitrary axes of coordinates, might make it useful in researches respecting the attractions of a system of bodies. A beginning of such a research had been made by Sir William Hamilton in October 1844...".

9. "The inner automorphisms of quaternions, $p \to qpq^{-1}$, where q is a quaternion of unit norm, $q\bar{q} = 1$, $\bar{q}$ being the conjugate of q, were used by Hamilton to give rotations in the space of the vectors of the quaternions" (H. Halberstam and R. Ingram in Hamilton, Vol. III, Appendix 1, 1967: p. 643).

"The more general mapping, $p \to a(p)b$, where a and b are quaternions each of unit norm, preserves invariant the norm of p. If p is written as $p = w + ix + jy + kz$ the square of the norm of p is $w^2 + x^2 + y^2 + z^2$ and this is invariant in four-dimensional Euclidean space (w, x, y, z). Thus there is a homomorphism from pairs of quaternions (a, b) to the rotation group in four dimensions" (*ibid.*).

10. As regards geometrical interpretation, Grassmann's approach (at least in the first edition of his book) is affine whereas Hamilton's is metric (cf. Klein, 1926, 1956: pp. 175, 187).

11. René Descartes contributed perhaps more than anyone else to the acceptance of the view that algebra (and eventually also analysis) was a symbolic *language* (rather than a science), convenient and precise in the formulation and effective in the solution of problems from other branches of mathematics, geometry for example, or from science. It was no coincidence that the fundamentals of algebraic (analytic) geometry were presented in an appendix ("Géométrie") *to Descartes's Discours de la méthode* (1637). In the *Discourse on Method* Descartes characterised the relational nature of mathematics (all the particular sciences commonly denominated Mathematics) as follows:
"... but observing that, however different their objects, they all agree in

considering only the various relations or proportions subsisting among those objects, I thought it best for my purpose to consider these proportions in the most general form possible, without referring them to any objects in particular, except such as would most facilitate the knowledge of them, and without by any means restricting them to these.... Perceiving further that in order to understand these relations... I should view them as subsisting between straight lines, than which I could find no objects more simple, or capable of being more distinctly represented to my imagination and senses; and on the other hand, that in order to retain them in memory, or embrace an aggregate of many, I should express them by certain characters the briefest possible. In this way I believed that I could borrow all that was best both in Geometrical Analysis and in Algebra, and correct all the defects of the one by the other..." (Part I).

In connection with the method of coordinates Descartes emphasised explicitly the role of free, arbitrary choices (of the origin of the coordinates of the unit-line, etc.), of the invariance of geometrical curves with respect to the transformation of coordinates and, hence, of the existence of equivalent descriptions. In *The Geometry of René Descartes*, (1637, 1954, Book II), entitled "On the nature of curved lines", we read:

"... if we think of geometry as the science which furnishes a general knowledge of the measurement of all bodies, then we have no more right to exclude the more complex curves than the simpler ones, provided they can be conceived of as described by a continuous motion or by several successive motions ... I think the best way to group together all such curves and then classify them in order, is by recognising the fact that all points of those curves which we may call "geometric", that is, those which admit of precise and exact measurement, must bear a definite relation to all points of a straight line, and that this relation must be expressed by means of a single equation.... If I want to find out to which class this curve belongs, I choose a straight line, as AB, to which to refer all its points, and in AB I choose a point A at which to begin the investigation. I say "choose this and that", because we are free to choose what we will, for while it is necessary to use care in the choice in order to make the equation as short and simple as possible, yet no matter what line I should take instead of AB the curve would always prove to be of the same class, a fact easily demonstrated..."

Most "geometers" (in the wide sense of the term) in the early nineteenth century acknowledged their indebtedness to Descartes (see the report on Hamilton's method in geometrical optics, in the present essay, as well as Plücker's and Lie's appreciation of the conventions implicit in Cartesian geometry from which further developments in ninteenth-century geometry ensued; see Chap. 1, this volume). However, the view that language, in particular mathematical language plays a much more fundamental role in cognition than the mere role of expressing and communicating our thoughts, seems to have originated in one form at least from Immanuel Kant's philosophy (see in this connection Chap. 4, this volume) and appears (perhaps independently of Kant) in Joseph Fourier (1822, 1878: pp. 6–8) and in George Boole (1854) which bears the traditional title of *An Investigation of the Laws of Thought* but contains an "algebra of logic" (of classes). Chap. 2 of the latter begins with the sentence: "That language is an instrument of human reason, and not merely a medium for the expression of thought, is a truth generally admitted...."

12. The most general linear transformation carrying the unit circle Q_o into itself and

carrying the interior of Q_o into itself is the transformation:

$$z' = \frac{az + \bar{c}}{cz + \bar{a}}, \quad c\bar{c} - a\bar{a} = 1$$

where $\bar{c}$ and $\bar{a}$ are conjugate imaginaries of c and a respectively (Ford, 1929, 1951: p. 31).

13. For the connection between differential equations and groups of linear transformations see (Ford, 1929, 1951: pp. 284–311).
14. An equivalent classification of theories into "principle theories" and "constructive theories" may be found in A. Einstein's 1919 article "What is the theory of relativity?" (reprinted in Einstein, 1954: pp. 227–32).
15. Poincaré claimed (1902, 1952: p. 166) that since one cannot give a general definition of or a unique interpretation to the term "energy", "... the principle of conservation of energy simply signifies that there is something which remains constant."
16. Felix Klein (1926: p. 215) used the term "mathematical physics" to apply to any phenomenological theory in physics which assumes a continuous medium and hence operates with partial differential equations. D'Abro (1939, 1951) distinguished mechanistic theories which postulate discrete or continuous media, field theories and phenomenological theories. Poincaré's "physics of central forces" may perhaps be identified with "mechanistic theories" but his "physics of the principles" is not identical with any of the categories used either by Klein or by d'Abro.
17. It seems that all these arguments in favour of replacement procedures have been ignored in recent disputes over Ramsey's and Craig's methods. Only semantical considerations arising from the criteria of the meaningfulness (or empirical interpretation) of theoretical terms have been taken into account both by those philosophers who defend and those who criticise Ramsey's or Craig's method (see, e.g. Putnam, 1965, 1979: pp. 228–36).
18. Though Ramsey himself seems to have thought that the class of the observational consequences of a theory is equivalent with (what we now call) its Ramsey–sentence, it was pointed out in more recent discussions of the subject that the latter is in fact logically stronger than the former, for the Ramsey–sentence of a theory excludes all those models of the observational part of the theory which cannot be expanded to the models of the full theory (Przełecki, 1974; Sneed, 1971).
19. If the mathematical form or structure of a theory is to be allowed to contribute to the theory's descriptive content, then one has to take into account the fact that there are forms or formalisms which are either equivalent or otherwise logically related. An interesting and most fruitful general theorem connecting variational, invariance and conservation principles was proved by Emmy Noether (1918) who combined for that purpose the methods of the variational calculus with Lie's theory of transformation groups. Roughly speaking, the theorem states that if the equations of motion are derivable from a variational principle (Hamilton's principle) and the variational integral is invariant with respect to a transformation group, then there exists at least one conservation theorem. Moreover, there is a systematic procedure for the establishment of such conservation theorems. For a discussion of the importance of this theorem see e.g. Kilmister (1970), Hill (1951). An interesting analysis of the epistemological and methodological status of various

types of principles in physics and of Noether's theorem in this context is contained in Nadel–Turonski (1976), where it is claimed that only invariance principles (in classical as well as in the "new" physics), the principles of correspondence and of complementarity in the old quantum theory and the principle of "micro-causality", in the relativistic quantum field theory are essentially meta-linguistic.
20. Klein (1926/7, 1956: p. 67) describes the situation in physics at the beginning of the nineteenth century in this way: "... there was an urgent need for the mathematician's ordering hand because of the commotion and confusion caused by new conceptions and theories refuting one another...."
21. This procedure is clearly different from what Putnam (1975, 1979: p. 161) calls "the conventionalist ploy" and Putnam's criticisms of the latter do not apply to it.
22. Let x, y, z be the rectangular coordinates of a point in Euclidean space and ξ, η, ζ analogous coordinates in Lobatchevskian space. A correspondence between the two spaces is then established by the following formulae:

$$\xi = \frac{2\alpha x}{x^2 + y^2 + z^2 - k^2}, \quad \eta = \frac{2\beta y}{x^2 + y^2 + z^2 - k^2},$$

$$\zeta = \frac{2\gamma}{x^2 + y^2 + z^2 - k^2}.$$

These transformation formulae were first given by G. Darboux in 1864 and then in his (1873) and (1894). Since projective formulae for distance and measure of angle have the same algebraic form in each of the three metric geometries, a one-to-one correspondence which preserves deductive relations may be established on this basis between the theorems of one geometry and the theorems of each of the others (Klein, 1928; Nagel, 1961: p. 248).

3

Duhem's Instrumentalism and its Critique: A Reappraisal*

THE MAIN objection against various empiricist and—in particular—positivist doctrines in nineteenth- and twentieth-century philosophy of science has been that they unduly restrict theorising: as descriptive accounts of natural and social science they refuse scientific status to various theories which many would normally count as science; as programmes for research they impose on theorising limitations which would hamper the development of certain theories and in this way—so some argue—stifle the progress of science. In addition, they brand as nonsense or—at least—as empirically meaningless various metaphysical speculations. Since apart from philosophers also many scientists have indulged in such speculations, it is felt that not only vital components of traditional philosophy but also of science have thus been ostracised. To be sure, very few of those empirically-minded scientists and philosophers of science wanted to dispense with or ban theories altogether and to restrict science to the "empirical basis" and empirical generalisations. Either because they felt that few scientists would want to be deprived of the "theoretician's paradise", or because they appreciated (with Poincaré) that "theoretical physics is a fact to be explained" or else because they saw the philosopher's role as descriptive rather than prescriptive, they have usually ruled against the ban on theories invoking an ancient clause in the epistemologist's legislature known in various forms as instrumentalism (formalism, anti-realism, etc.). The intended effect of that clause is that those theoretical terms and sentences which do not satisfy certain definite requirements in terms of relations to observables (translatability into observables in the extreme case of phenomenalism, descriptivism or positivism; increase in predictive power of the theory in the case of more liberal empiricist doctrines) are acknowledged as components of scientific theories with the status of formal symbols, metaphors, useful fictions, instruments, etc. In contemporary social sciences (economics, in particular) the instrumentalist clause would allow so-called "models" apart from proper empirical theories as well as theories based on unrealistic assumptions.[1] This piece of liberal legislation has not been appreciated at all by recent antipositivist critics of those empiricist doctrines, for it seems to those critics that it does not take theorising sufficiently seriously and, consequently, grants theoreticians too great freedom blurring the boundary between science and fiction. Thus modern empiricism has been blamed by its various critics for being intolerant and at the same time too permissive.

The aim of the present essay is to re-examine some recent accounts and critical assessments of the instrumentalist tradition in natural science and its philosophy. I shall claim that those accounts are descriptively inadequate. They are inadequate because they represent instrumentalism exclusively in ontological or semantical terms, viz. either as a doctrine which refuses existence to unobservable entities or as a doctrine which regards theoretical terms as uninterpreted. Although both these views have played some role in the instrumentalist tradition in science and its philosophy, mainly as a reaction to naïve realism, there are two other important components without which that tradition cannot be properly understood: the pragmatic and the epistemological components. The pragmatic component is concerned with the question under what conditions it is rational to *assert* a theory and when should one *suspend judgement* and merely entertain the theory. The epistemological component is concerned with finding the *limits* of knowledge determined either by practical circumstances (available experimental and measuring techniques) or—more importantly—by permanent factors such as observational equivalence of theories, isomorphism of postulated unobservable structures, etc. In the light of this there is a need for revising both our concepts of realism and instrumentalism and for a revaluation of their role in science and philosophy.

I shall first report two accounts and assessments of instrumentalism given by Karl Popper and Paul Feyerabend, respectively. Then I shall try to show their inadequacy with respect to the instrumentalist tradition in astronomy and with respect to nineteenth-century conventionalism.

Two Views Concerning Instrumentalism and its Effect on Natural Science

Perhaps the earliest and strongest condemnation of instrumentalism in contemporary philosophy of science came from Sir Karl Popper. It was made in the context of the philosophy of the natural sciences, primarily of physics and astronomy.

Popper contrasted three views concerning the status of physical theories, essentialism, instrumentalism and "the third view", realism combined with fallibilism (hypotheticism), which is also Popper's own view (1963: pp. 97–120).

Essentialism, which originated in Aristotle's philosophy and shared its age-old influence on science and philosophy, asserts that (a) the aim of science is to give a true description of the world, (b) it is possible for science and philosophy to demonstrate which theories about the world are true, (c) scientific and philosophic theories about the world concern the hidden nature of things, essential properties, true causes by contrast to observable phenomena, (d) explanations of phenomena in terms of latent causes, natures or essences are ultimate, i.e. in no need of further explanation.

From the third point of view, i.e. Popper's own, claims (a) and (c) are descriptively correct and programmatically healthy. Claims (b) and (d), on the other hand, are mistaken and unacceptable. It is, however, the last of the claims which Popper blamed as obscurantist, since by preventing scientists and philosophers from asking fruitful questions concerning deeper and deeper explanations, it hampered the progress of science.

Instrumentalism was identified by Popper with the view concerning the status of scientific theories "... founded by Osiander, Cardinal Bellarmino and Bishop Berkeley..." but also advocated "in various ways" by Mach, Kirchhoff, Hertz, Duhem, Poincaré, Bridgman, Eddington as well as by Niels Bohr and those who have adopted the so-called Copenhagen interpretation of quantum physics (1963: p. 99). The following tenets are, according to Popper, characteristic of instrumentalism.

Scientific laws and theories are not proper, descriptive statements; they are nothing but instruments to derive observational predictions from other observational statements. Dispositional and theoretical terms (e.g. magnetic, force, field of forces), unlike observational terms, do not refer to real entities but are merely symbols which facilitate deductions. When a theory is found not to work in a given area, its field of application is suitably modified.

Since scientific theories are nothing but computational rules, of the same character as the computational rules of applied science, there is no difference between pure and applied science; all science is applied. By contrast, realism combined with fallibilism claims that: scientific laws and theories are not only instruments for making predictions but are also genuine, descriptive (true or false) statements; they are empirical hypotheses which have explanatory power. Dispositional and theoretical terms refer to real entities just as so-called observational terms do. All terms are theoretical to some degree, though some are more so than others; the distinction, therefore, between observational and theoretical terms, assumed by instrumentalists, is mistaken.

Scientific theories, though not verifiable by observation, can clash with reality, provided they are falsifiable; they do clash, if—when subjected to genuine tests, i.e. attempts at falsification—they are shown to be falsified by experimental outcomes; crucial experiments, though unable to establish any theory, can eliminate some as false.

The fundamental aim of theoretical (pure) science is to provide a (hopefully) true description of the world, explanations of increasing depth (which means, of increasing abstractness, referring to unobservables, more and more removed from everyday experience) and predictions, some of which at least are predictions of new kind of events, i.e. are real discoveries (Popper, 1963: pp. 107–10, 114–19).

There are several reasons why, according to Popper, instrumentalism gained popularity among scientists and philosophers. One of those reasons was the fact that it enabled them "to deal with inconvenient scientific hypotheses". So, for example, it enabled Osiander and Bellarmino to deal with the Copernican hypothesis in such a way that it no longer appeared to contradict the Scripture. Another reason, related to the previous one, was the concern that if science is credited with the ability to discover the truth about the world unaided by divine revelation, religion would suffer. This was, for example, Berkeley's concern. There was a third reason. Many scientists and philosophers rejected the essentialist view that the aim of science was to discover hidden causes or essences of things, either because they did not believe that such hidden causes or essences existed (e.g. Mach) or because they did not believe science was able with its methods to discover them (e.g. Berkeley, Duhem, Poincaré). Consequently they denied that theoretical concepts referred to latent causes or properties and that the aim of science was to give explanations which transcend our knowledge of the phenomenal world. Similarly, from the correct belief that science is unable to demonstrate the truth of its theories and that so-called crucial experiments cannot establish any theory as true, instrumentalists (e.g. Osiander, Bellarmino, Duhem, Poincaré) drew the wrong conclusion that scientific theories are never descriptive, possibly true statements but are merely mathematical tools for making predictions. Similarly, from the (right) premiss that the method of science cannot yield but tentative, revisable hypotheses and from the further premiss that scientific laws and theories (e.g. Newton's laws of motion, conservation laws) cannot be contradicted by experimental findings nor can a crucial experiment be instituted to refute, e.g. one (physical) geometry in competition with another, some instrumentalists (e.g. Poincaré) drew the wrong conclusion that scientific laws and theories are nothing but terminological conventions, useful for systematising observational data in the way a catalogue systematises books in a library, but having no descriptive, empirical content (Popper, 1963: p. 98).

Paul Feyerabend gives two related characterisations of instrumentalism and realism. In an earlier article (Feyerabend, 1964: pp. 280–309) he defined "instrumentalism" in a way similar to Popper's, except that—in conformity with the traditional usage established in the history of astronomy—he does not mention fallibilism (hypotheticism) in formulating the position of realism. In a later article (Feyerabend, 1970: p. 220) he associates instrumentalism with the claim that there is a theory-independent observational language available for making comparisons between rival theories and realism with the denial of this claim (cf. the same denial on Popper's list of the tenets of the "third view") as well as with the practice of using "the most abstract terms" of relevant theories to give theoretical meaning to all observables. In effect, an instrumentalist "makes commensurable all those theories which are related to the same observation language and are interpreted on its basis". Not so the realist: theoretical reinterpretation of observables in the light of a new theory may be so radical that they disappear from the class of consequences shared with earlier theories; the result is incommensurability (Feyerabend, 1970: p. 220).

In his earlier treatment of instrumentalism Feyerabend pointed out that in Copernicus's theory three types of argument were used in support of instrumentalism: religious (the inconsistency of the realist version of the theory with the Scripture); epistemological (only theories whose truth has been conclusively demonstrated can be descriptive of physical reality); and physical (inconsistency with Aristotelian physics—well-confirmed, given available evidence). In a somewhat "naive falsificationist" mood, Feyerabend contended that a "definitive refutation" of the epistemological arguments (of Bellarmino and Duhem) may be found in Popper's article on the three views. He therefore concentrated on the physical arguments—in order to show that, even if these (according to him, weightier) arguments are taken into account, realism is preferable to instrumentalism. Now, the examination of the physical arguments in support of instrumentalism show—so Feyerabend continued—that one has to revise drastically traditional methodology which demanded factual support of new theories as a condition of their acceptability. Since certain defects of established theories cannot be detected without the help of alternative theories, one must not wait for the established theory to be directly refuted by experimental findings but proliferate rival theories, inconsistent with the established one, take them in the strongest form, i.e. in the realist fashion and—if any of them is more successful in solving problems or yielding new predictions—then the old theory, though unrefuted directly, ought to be replaced by the new, more successful one (Feyerabend, 1964: p. 307). This methodological principle of proliferation of mutually inconsistent and incommensurable theories (cf. Feyerabend's second characterisation of realism) is offered by Feyerabend as a "methodological justification for realism" and claimed by him to be demanded by "the principle of testability, according to which it is the task of the scientist relentlessly to test whatever theory he possesses" (Feyerabend, 1964: p. 305).

As stated before, I believe that both Popper's and Feyerabend's accounts of the instrumentalist tradition are historically inadequate and their negative evaluation of that tradition unjustified. Since, however, some representatives of instrumentalism in the Popper–Feyerabend sense may be found in the history of science and philosophy, perhaps it would be advisable to distinguish varieties of instrumentalism and realism. This could be done by distinguishing, on the one hand, between ontological, semantical, pragmatic and epistemological theses (or senses) of instrumentalism and between strong and moderate versions of the doctrine, depending on which and how many of the mentioned theses are involved.

As regards Popper's strong condemnation of instrumentalism, it should be pointed out that he compared the instrumentalist tradition with his own philosophy ("the third view"), i.e. realism combined with fallibilism. Historically instrumentalism was never opposed to realism in this sense but to naïve realism. Fallibilism (hypotheticism) or even scepticism has almost always been the monopoly of instrumentalists. A fair appraisal of the historical role of the instrumentalist tradition must take this into account.

The Instrumentalism of Osiander and Bellarmino

Without attempting to solve the historical problem of the genesis of instrumentalism (or of anti-naïve-realism) I shall assume here as uncontentious that the views concerning the aims and status of astronomical theories expressed by Osiander and Bellarmino did not originate with those two authors but were the result of a long tradition going through the Middle Ages back to ancient Greek astronomy. They originated directly from the distinction established by ancient Greeks between mathematical and physical astronomy. As Geminus put it (in the first century B.C.), the business of a physicist is "to consider the substance of the heavens and stars, their force and quality, their coming into being and destruction.... The physicist will prove each fact by considerations of essence and substance ... (he) ... will in many cases reach the cause by looking to creative force." Not so the astronomer who is interested in "the shapes and sizes and distances of the earth, sun and moon, and of eclipses and conjunction of the stars"; accordingly he makes use of "arithmetic and geometry" and "is not qualified to judge of the cause ... sometimes he does not even desire to ascertain the cause, as when he discourses about an eclipse; at other times he invents by way of hypothesis and states certain expedients by assumption of which the phenomena will be saved." The task of mathematical astronomy, accordingly, is to invent kinematic systems capable of saving the phenomena "... without being concerned with the question whether such systems are realized in the physical structure of the heavens, and if so, in what way..." (Geminus in Cohen and Drabkin, 1969: pp. 90–1). The distinction, apparently influenced by Aristotle's conception of physics, does not imply instrumentalism in Popper's sense, i.e. the claim that astronomical hypotheses are not descriptive, true or false sentences. All that is implied by it is that *mathematical astronomers are entitled to leave the question of the physical realisations of their mathematical systems open and that—as astronomers— they are not qualified to do otherwise.* This was repeated almost verbatim by Osiander in his preface to Copernicus's *De Revolutionibus*:

> ... For it is the duty of an astronomer to compose the history of the celestial motions through careful and skilful observation. Then turning to the causes of these motions or hypotheses about them, he must conceive and devise, since he cannot in any way attain to the true causes, such hypotheses as, being assumed, enable the motions to be calculated correctly from the principles of geometry, for the future as well as for the past. The present author has performed both these duties excellently. For these hypotheses *need not be true* nor even probable; if they provide a calculus consistent with the observations, that alone is sufficient... (Osiander in Rosen, 1969: pp. 24–5).

The point that although astronomical hypotheses *may be true or false* the astronomer does not commit himself to the belief in either, is made even more clearly in Osiander's letter to Copernicus:

> ... I have always felt about hypotheses that they are not articles of faith but the basis of computation; so that *even if they are false* it does not matter, provided that they reproduce exactly the phenomena of the motion ... (Rosen, 1969: p. 22).

Similarly, and even more cogently, Bellarmino wrote in his letter to Foscarini:

> ... To say that on the supposition that the earth moves and the sun stands still all the appearances are saved better than on the assumption of eccentrics and epicycles, is to say very well—there is no danger in that, and it is *sufficient for the mathematician*: but to wish to *affirm* that in reality the sun stands still in the center of the world, and only revolves upon itself without travelling from east to west, and that the earth is located in the third heaven and revolves with greater velocity about the sun, is a thing in which there is much danger not only of irritating all the scholastic philosophers and theologians, but also of injuring the Holy Faith by rendering false the Sacred Scriptures.... I say that *once there is a real demonstration* that the sun stands in the center of the world and the earth in the third heaven, and that the sun does not go round the earth, but the earth around the sun, then we must go to work with much thoughtful consideration to explain the passages of Scripture that seem to oppose this view, and better to say that we have not understood these passages, than to say that that which has been demonstrated is false. But I shall not believe that there is any such demonstration until it has been shown to me: nor is it the same thing to demonstrate that supposing the sun to stand in the center and the earth to move in the heaven will save the appearances, and to demonstrate that in truth the sun does stand in the center and the earth moves in the heaven; for I believe that the first demonstration can be given, but concerning the second I have the gravest doubt, and *in case of doubt* one should not abandon the Sacred Scriptures as they have been expounded by the Holy Fathers (Bellarmino in Duhem, 1969: p. 107).

This seems to be the passage in Bellarmino's letter which Popper construes as concluding the non-descriptive nature of the hypothesis of the earth's motion (instrumentalism in Popper's sense) from the fact that the truth of the hypothesis has not been demonstrated and it is this alleged epistemological argument for instrumentalism, along with a similar Duhemian one, which Feyerabend believes to have been definitely refuted by Popper. But what Bellarmino explicitly says in this passage are the following points: (a) the truth of the hypothesis of the earth's rotation and revolution has not been demonstrated, for (b) to show that the hypothesis explains (saves) all relevant phenomena is not (the same as) to demonstrate the truth of the hypothesis, therefore (c) until such demonstration is presented, one should *abstain from affirming its truth* and be satisfied with using it *ex suppositione* (*ex hypothesi*), (d) this is, in fact, sufficient for the mathematical astronomer, and (e) resolves for the time being (i.e. until a demonstration of the hypothesis is given) the clash with the Scripture and with peripatetic physics.[2]

Clearly, to advise not to assert a hypothesis as true but merely entertain it—which is what Bellarmino does—is not the same as to advise to divest it of its physical interpretation. Logical inconsistency with the Scripture and with peripatetic physics may be avoided not only through deinterpretation but also by *suspending assertion* of one of the mutually inconsistent statements. But suspension of assertion and mere entertainment were essential from the anti-naïve-realist viewpoint in astronomy since proliferation of "factually adequate and mutually inconsistent theories" or "invention of alternatives in addition to the view that stands in the centre of discussions" (Feyerabend, 1968: p. 29) were no less strongly emphasised by its methodology than by Feyerabend's, although for somewhat different reasons. Here is Osiander's plea for tolerance in matters astronomical:

> ...The peripatetics and theologians will be readily placated if they hear that there can be different hypotheses for the same apparent motion; that the present hypotheses are brought forward, not because they are in reality true, but because they regulate the computation of the apparent and combined motion as conveniently as may be; that it is possible for someone else to devise different hypotheses; that one may conceive a suitable system, and another a more suitable, while both systems produce the same phenomena of motion; that each and every man is at liberty to devise more convenient hypotheses; and that if he succeeds, he is to be congratulated. In this way they will be diverted from stern defense and attracted by the charm of the inquiry; first their antagonism will disappear, then they will seek the truth in vain by their own devices, and go over to the opinion of the author (Rosen, 1959: p. 23).

In his own way, Osiander argued that progress in astronomy may be achieved better through tolerance by letting everyone invent his alternative to the existing theories, through greater simplicity (given, however, the same empirical adequacy of alternatives), by avoiding antagonism and attracting opponents through "the charm of the inquiry", in other words—replacing the old theory by better ones without any of the theories being either conclusively refuted or demonstrated. In effect, this is Osiander's methodological justification of the anti-naïve-realist view.

Duhem's most often discussed and criticised comments on the controversy between realists and antirealists in Renaissance astronomy were made in connection with Galileo's indirect retort to Bellarmino. In a statement written towards the end of 1615 and addressed to the consultants of the Holy Office, Galileo first appeared to grant Bellarmino that one must not expect anyone to believe in the mobility of the earth without a demonstration; then he also conceded that "... it is not the same thing to show that on the assumption of the sun's fixity and the earth's mobility the appearances are saved and to demonstrate that such hypotheses are really true in nature...."; finally, Galileo postulated:

> ... But it should also be granted, and is much more true, that on the commonly accepted system there is no accounting for these appearances, whence this system is indubitably false; just so should it be granted that a system that agrees very closely with appearances may be true; and one neither can nor should look for other or greater truth in a theory than this, that it answers to all the particular appearances (Galileo in Duhem, 1969: p. 108).

In the quoted passage Galileo appears to have distinguished—as was customary at the time—two senses of "true", viz. "true in nature" and "saving all the appearances". He also distinguished two methods of demonstration, one appropriate to establishing truth in nature, the other able to eliminate the hypotheses which do not account for the appearances and to establish that a hypothesis does save the phenomena. He seems to have claimed that only the second kind of truth may be sought in scientific hypotheses and that, therefore, the method of science need not be a demonstration of the first kind, in fact it could only be of the second kind. We do not know, of course, whether these statements conveyed Galileo's own views. Most likely they constitute a compromise, clothed intentionally in somewhat ambiguous expressions, between his views and those of his critics, for they do not appear to be more in the spirit of realism than those of Osiander or Bellarmino.

Admittedly it is also possible to read Galileo's words with an emphasis on the idea of a crucial experiment, which is implicit in them. This is apparently how Duhem chose to see their import and in doing so he read into the passage the claim—not explicitly expressed by Galileo—that the experimental method of science conveys certainty and truth on one of two hypotheses which survives the confrontation with its rival and with the appearances:

> ...Galileo's notions of the validity of the experimental method and the art of using it are nearly those that Bacon was later to formulate. Galileo conceives of the proof of a hypothesis in imitation of the *reductio ad absurdum* proofs that are used in geometry. Experience, by convicting one system of error, confers certainty on its opposite. Experimental science advances by a series of dilemmas, each resolved by an *experimentum crucis* (Duhem, 1969: p. 109).

To show the inadequacy of Galileo's account of experimental method Duhem formulated his celebrated argument against the possibility of crucial experiments:

> ...Grant that the phenomena are no longer saved by Ptolemy's system; the *falsity* of that system must then be acknowledged. But from this it does not by any means follow that the system of Copernicus is true; the latter is, after all, not purely and simply the contradictory of the Ptolemaic system. Grant that the hypotheses of Copernicus manage to save all the known phenomena; that these hypotheses *may be true* is a warranted conclusion, not that they are *assuredly true*. Justification of this last proposition would require that one prove that no other set of hypotheses could possibly be conjured up that would do as well at saving the phenomena. The latter proof has never been given. Indeed, was it not possible, in Galileo's own time, to save all the appearances that could be mustered in favour of the Copernican system by the system of Tycho Brahe? (Duhem, 1969: pp. 109–10).

According to Popper (also Feyerabend and Rosen), Duhem's argument merely shows that there are no conclusive crucial experiments, i.e. experiments which by refuting one of two rival (contrary but not mutually contradictory) hypotheses would conclusively prove the truth of the other. But the impossibility of conclusive crucial experiments does *not* imply—contrary to Duhem—the instrumentalist (uninterpreted, purely formal) status of all scientific hypotheses.

This criticism does not seem to be quite fair. Moreover, it misses the most important point in Duhem's comments, which—to my mind—is not only the claim that crucial experiments are inconclusive (owing to the fact that rival hypotheses are usually contraries rather than contradictories, i.e. two such hypotheses cannot both be true but they can both be false) but also the claim that for any set of observed phenomena there are always several—possibly, infinitely many—observationally adequate or observationally equivalent and adequate, conceptually distinct hypotheses. As a consequence, it is impossible ever to single out with the help of the experimental method of science alone *one* observationally adequate hypothesis as possibly true; *with any one such hypothesis we have always many*. This is a limitation of the experimental method of science (or of its "scope" as Duhem put it) which Galileo and other realists seem to have ignored. However, owing to this limitation we can at most claim of a hypothesis that it belongs to *a class* of those hypotheses which save all the phenomena in the given domain of investigation equally well. If for any set of data there are always many empirically adequate hypotheses and, more importantly, if for any empirically adequate hypothesis there are always many others observationally equivalent with the original one, then either all such hypotheses would have to be regarded as synonymous—in which case "true" would simply mean "observationally true"—or else, among observationally equivalent hypotheses there would be some mutually incompatible ones—in which case all the relevant empirical evidence (not just evidence available at the moment), i.e. the only experimental ground for judging a hypothesis as possibly true, would invariably support two or more mutually incompatible hypotheses. Should one want to affirm (assert) hypotheses on experimental grounds, then one would have to affirm on the basis of the same evidence mutually incompatible hypotheses. It was to avoid this apparent paradox of experimental method[3] as well as to emphasise the extremely tentative and transient nature of its conclusions, that Duhem believed one should regard hypotheses as merely *mathematical contrivances or instruments*, i.e. as neither true nor false "in nature" (but as true or false in the sense of saving or not saving the phenomena) and that one should abstain from affirming them as such:

The physicists of today... having seen so many illusions dissipated that previously passed for certainties, have been compelled to acknowledge and proclaim that logic sides with Osiander, Bellarmino and Urban VIII, not with Kepler and Galileo.... we believe today, with Osiander and Bellarmino, that the hypotheses of physics are mere mathematical contrivances devised for the purposes of saving the phenomena. But thanks to Kepler and Galileo, we now require that they save all the phenomena of the inanimate universe together (Duhem, 1969: pp. 113, 117).[4]

The Instrumentalism of Duhem

The title of one of the sections of Duhem's article "Physical law" (in *The Aim and Structure of Physical Theory*) announces that "a law of physics is, properly speaking, neither true nor false...." (1906, 1968: p. 302). What clearer evidence could there be of Duhem's instrumentalism? Is this not enough to disregard, as his critics have done, Duhem's apparent admission in *To Save The Phenomena* of the descriptive (true or false) nature of rival astronomical systems? I shall consider now these questions in relation to another version of the Duhemian argument against the possibility of crucial experiments, viz. the argument from the existence of experimentally indistinguishable hypotheses.

Of course, Duhem's statement concerning the laws of physics appears to be a clear indication of his instrumentalism only as long as it remains—as quoted above—in incomplete form. When completed it goes as follows: "A law of physics is, properly speaking, neither true nor false but approximate."

One does not refer to mere instruments, which cannot perform a descriptive function as "approximate". And indeed from Duhem's elucidation of what he means by "approximate" it becomes obvious that he regards physical laws as descriptive, although not in the same, simple way as "common-sense laws" (1906, 1968: pp. 299–306). With regard to commonsense laws, which are simply generalizations of everyday experiences, e.g. "In Paris, the sun rises every day in the east, goes up in the heavens, then comes down and sets in the west", the question whether they are true or false is determinate and can be answered with a "yes" or "no". Not so with respect to scientific laws which are symbolic representations picturing "... reality in a more or less precise and detailed manner." *The question whether such symbolic representations of reality are true or false is indeterminate and must not be asked.* The reason is this: a physical law is always underdetermined by the evidence we possess, in this sense it may be said to be indeterminate; a given "group of facts" (evidence statements) is compatible with an infinity of formulae which may express our law. As an example, let us consider a physical law corresponding to the aforementioned common-sense law. Such a physical law would be a function which assigns to each moment of time the position of the sun in the sky as seen from Paris. Of course the law would be concerned not with the sun as we see it but with the positions of the centre of the sun conceived as a geometrically perfect sphere disregarding atmospheric refraction and annual aberrations; this is one of the senses in which a physical law is symbolic, i.e. it is a redescription of facts in the idealised and precise language of mathematics. To make this symbol "correspond to reality" we have to effect complicated measurements, using a lens equipped with a micrometer, make many readings, subject these to corrections and calculations, the legitimacy of which depends on various theories, e.g. on the theory of aberrations and of atmospheric refraction. Our formulae tell us the coordinates of the point we call the centre of the sun; the meaning of these cannot be understood without knowing the laws of cosmography and their values designate points in the sky only given various previous determinations (e.g. of the meridian of the place, etc.). Now, owing to *limited sensitivity of our optical instruments*, we shall be able to determine for each instant the longitude and the latitude of the centre of the sun only with a limited precision, e.g. not greater than $1'$; we shall not be able to discriminate between two positions which differ from one another by less than that value. Hence, in spite of the fact that the sun's centre occupies at each instant of time only one position, we shall be able to give for each instant an infinity of values for the longitude and an infinity for the latitude. Consequently "in order to represent the path of the longitude as a function of the time, we shall be able to adopt not a single formula, but an infinity of different formulas, provided that for a given instant all these formulas give us values for the longitude differing by less than $1'$." And the same applies, of course, to the latitude. We shall, then, be able to represent our observations of the path of the sun by an infinity of different laws, expressed by equations which are mutually incompatible, i.e. if one of them is true, no other is. Each of these equations will trace a different curve on the celestial sphere:

> ... It would be absurd to say that the same point describes two of these curves at the same time, yet, to the physicist all these laws are equally acceptable, for all determine the position of the sun with a closer approximation than can be observed with our instruments. The physicist does not have the right to say that any of these laws is true to the exclusion of the others... (Duhem, 1906, 1968: p. 305).

The physicist will eventually choose one of the laws but his considerations will not be necessarily related to the idea of truth and they may vary from case to case depending, for example, on the use to which the law is to be put. Among the considerations which guide his choice may be simplicity but especially the deducibility of the given law from a theory which he had accepted; however "... physical theories are only a means of classifying and bringing together the approximate laws to which experiments are subject; theories, therefore, cannot modify the nature of these experimental laws and cannot confer absolute truth on them..." (1906, 1968: pp. 305–6).

The degree of the indetermination of abstract symbols (theoretical concepts, as we would say today) depends on the degree of approximation with which quantities can be measured. However, the latter varies with time and increases with the improvements of instruments. Accordingly, what was acceptable as sufficiently precise at one time, will not be so acceptable at a later time. On the other hand, the degree of precision required varies with purpose. Moreover the same physical law is often simultaneously accepted and rejected by the same physicist in the course of the same work; this would result in a formal contradiction "if a law of physics could be said to be true or false". Finally, a physical law is in the process of constant modifications and improvements; factors previously disregarded are being taken into account, exceptions are explained which had only been enumerated before, etc. As a result, it is more a series of statements than one complete statement.

There are at least two distinct ideas in Duhem's argument. Firstly, he points out that a set of results of physical measurements made to test a physical law formulated in the precise language of analysis always supports—on account of limited precision of physical measurement—not only the one specific law which we intended to test, but an infinite number of mathematically distinct but experimentally indistinguishable laws. If a choice *is* made between them, then, naturally, this choice is not based on experimental grounds but on the basis of some other, pragmatic considerations such as simplicity in some sense or preference for some theory into which one law fits better than another. Secondly, Duhem claims that a physical law—expressed in a mathematical formula, e.g. $f = m.a$—changes its meaning and epistemological status over time when pragmatic factors are taken into account, such as improved measuring techniques or different research and practical aims, etc. The result is that we deal with a matrix or a sequence of statements which are not consistently accepted (asserted) or rejected by the scientific community or even by one and the same scientist. Duhem concludes from both these claims that laws must not be regarded as true or false but rather as indeterminate statements since otherwise science or a theory of which they are part would not be a consistent system.

The least controversial seems to be the claim that a scientific law is not uniquely imposed upon us by the results of experiments, however precise; decisional elements are involved in scientific procedures. Most people would agree today, so it seems, that this is a valid criticism of and an improvement on the traditional empiricist account of science.

If one decides to regard a mathematical formula combined with different pragmatic applications, etc. as *one* and the same law with variable content, then naturally it would not be advisable or indeed possible to assign to it in a consistent reconstruction one of the two truth-values of classical logic. A law as a sequence (possibly unended, open) of statements is not a unit which can be handled within ordinary logic. The suggestion that a law in this sense is neither true nor false (in the classical sense) is a rather obvious one and should not be a basis for the distinction between the instrumentalist and the realist view of scientific theories.

As regards the problem of the epistemological status of experimentally indistinguishable laws formulated in precise mathematical language (with precise values of constants), one must point out first of all that it arises only with respect to those laws which remain so indistinguishable, however much the precision of physical measurements may be improved. Let us refer to them as indistinguishable in principle. Moreover, this specific problem would be easily solved if in principle experimentally indistinguishable laws were regarded as observationally equivalent and the latter as equivalent. For then all mathematically precise (idealised) in principle experimentally indistinguishable laws would be equivalent, hence *not* mutually incompatible. Like other theoretical, observationally equivalent, hypotheses, they would be regarded as merely linguistic variants of the same law or as different (mathematical) metaphors expressing the same law. The necessary and sufficient condition for such a law to be true would be that all its observational consequences should be true. Precise, in principle experimentally indistinguishable, laws would in this way have the status of true or false statements. The identification of "observational equivalence" with "equivalence" on which this solution is based would—according to some—in itself be enough to classify the solution as instrumentalist. Nevertheless, they might still find some satisfaction in being able to assign a truth-value to laws. However, the solution mentioned would, of course, not avoid another difficulty to which Duhem and Charles Peirce drew our attention and which was later to be emphasized so strongly by Quine, viz. that it seems impossible to identify an individual law's observational consequences in isolation from other laws. Should this be an intractable difficulty, then the escape would be sought in several directions. One of these is instrumentalism and conventionalism, another a holistic view of science obviously inspired by conventionalism and pragmatism. Another possibility might be the recently much explored structuralist (non-statement) view of theories. A purely historical or socio-psychological study of science clearly avoids this specific epistemological problem since it avoids all epistemology in the traditional sense.

To conclude:

Duhem's instrumentalism was not based on an invalid argument from the inconclusive nature of crucial experiments to the instrumentalist status of hypotheses and laws. Its basis was—as we have seen—the quite plausible claim that there are observationally equivalent and experimentally (in principle) indistinguishable statements some of which—if allowed to have truth-values—would be mutually incompatible and would have exactly the same experimental support. This Duhem saw—as did Poincaré—as an important limitation of the experimental method of science. He differed from Poincaré[5] in that he decided to regard observational equivalence (or experimental indistinguishability in principle) as sufficient for two theories to be simply equivalent or even synonymous and in that he believed that there was another kind of knowledge, viz. metaphysical or religious, which was not subject to the same limitation since it did not make use of the experimental method.

Notes

[*] First published in "Essays in memory of I. Lakatos", *Boston Studies in The Philosophy of Science*, **XXXIX**, D. Reidel Publishing Co. 179–207, (1976). The last two sections have been abridged.
1. Cf. the dispute originated by M. Friedman (1963).
2. By the time Bellarmino wrote his letter to Foscarini in 1615, leading Jesuit astronomers, on whose opinion Bellarmino relied, were convinced by Galileo's and their own discoveries, that the Ptolemaic system did not save the phenomena. Since there were both religious and physical arguments (e.g. failure to discover stellar parallax) against the Copernican hypothesis, they were in favour of the Tychonic system.
3. Duhem's critique of the traditional conception of the "scope" of the experimental method anticipates N. Goodman's "new riddle of induction" and the "grue-emeralds" paradox. Cf. N. Goodman *Fact, Fiction and Forecast* (1965).
4. In the third of the quoted passages Duhem explicitly admits that the Ptolemaic and Copernican Theories are true or false, or rather that given suitable evidence one can show that one of them is definitely false and the other *may* be true. This seems like a straightforward rejection of instrumentalism in Popper's sense rather than an argument in its favour. One would avoid this difficulty by assuming that "true" means here "observationally true".
5. For Poincaré two theories to be synonymous (to have the same descriptive content) have to be not only observationally (numerically) equivalent but also to have a mathematical structure (form) of the same type. For a recent attempt to clarify the concept of "structure of the same kind" see Scheibe (1979).

4

Radical Conventionalism, its Background and Evolution: Poincaré, LeRoy and Ajdukiewicz*

I. The Evolution of Ajdukiewicz's Philosophy

THE DOMINANT theme of Ajdukiewicz's philosophy throughout his life was the problem of the dependence of our knowledge and of our conception of knowledge on language. Two related questions, on two different levels, were involved: one was the question whether our scientific world-view may be uniquely determined by experience or whether it is rather co-determined by the choice of a language; the other is the question whether the solutions of the fundamental epistemological problems—in particular our conception of knowledge—are independent of the choice of a conception of language. Epistemology, philosophy of science, of mathematics and of language were the areas in which Ajdukiewicz's philosophical interests were concentrated and to which he made the most important contributions.

The philosophy which emerged from his preoccupation with two main epistemological problems mentioned, underwent within over forty years of his more mature intellectual life an interesting, though perhaps not unusual, evolution.

In the *first period*, roughly before 1936, it was dominated by the doctrine which Ajdukiewicz himself labelled *radical conventionalism*. Radical conventionalism resulted from an original pragmatic conception of language combined with a concept of intersubjective meaning which Ajdukiewicz developed between 1929 and 1934 ("On the meaning of expressions", "Language and meaning"). The doctrine affirmed: (1) the existence of languages or conceptual frameworks which are not intertranslatable; (2) the necessity of articulating any knowledge in one of those languages; (3) the possibility of choosing one of the languages or of changing from one to another, hence of the existence of a decisional or conventional element in all knowledge; (4) the discontinuous nature of changes in science throughout its history. It is not difficult to see in radical conventionalism a continuation of the post-Kantian conventionalist tradition and also an anticipation (by some thirty years) of many of the current doctrines in contemporary philosophy of science.

Radical conventionalism, conceived and developed at the time when logical positivism in Central Europe reached its climax, was presented to the philosophical world outside Poland in a series of articles in *Erkenntnis* in 1934–5 ("Language and meaning", "The world-picture and the conceptual apparatus", "The scientific world-perspective"). Like the philosophy of the Vienna Circle and the Berlin Circle, it was inspired by various developments in mathematics and in science which occurred during the nineteenth century and in the first two decades of the present century, viz. the discovery of non-Euclidean geometries, the crisis in the foundations of mathematics (set-theoretic paradoxes) and various methods of resolving the crisis, the development of mathematical logic (in particular of meta-logic), the transition from classical to relativistic and quantum physics, etc. However, unlike the logical positivists of the Vienna Circle, Ajdukiewicz was never under the influence of phenomenalism (of either Berkeley's, Hume's or Mach's type); nor did he dismiss as absolutely meaningless all the problems and theses of traditional metaphysics. So, for example, in his critique of the reistic analysis of the problem and doctrine of universals he argued in 1935 ("On the problem of universals") that it is illegitimate to "translate" a metaphysical thesis formulated in one language (in this case of Aristotle) into another language (e.g. conforming to Kotarbinski's reistic criteria) and to criticise the "translation" as if it were the original thesis; it is illegitimate to do so because the languages of different philosophical systems are not mutually translatable.

On the whole, the doctrine of radical conventionalism had a much wider—hence, more liberal—philosophical background than its contemporary, logical positivism. Ajdukiewicz himself saw radical conventionalism as a critical, revised (radicalised) continuation of the philosophy of the new critique of science (*la nouvelle critique de science*) in France, in particular of Henri Poincaré and Edouard LeRoy. He emphasised strongly the Kantian features of his doctrine, as well as the indebtedness of his philosophy of language to some ideas of Bolzano and Husserl. Finally, he appreciated and endorsed the hermeneutic idea of understanding (Dilthey, Spranger) as the method—or at least as one of the methods—of the humanities, including epistemology and the history of philosophy. The latter view was never abandoned by Ajdukiewicz.

The retreat from radical conventionalism was first acknowledged at a public discussion in 1936 and, again, many years later, in a polemical article.[1]

However, it seems that its fundamental ideas never ceased to fascinate Ajdukiewicz. A critical re-thinking of those ideas was certainly responsible for some of the most original articles produced during the *second period* of his life. In particular, it was responsible for his clarification and defence—against his own arguments put forward during the radical conventionalism period—of the thesis of radical empiricism, according to which all knowledge consists of empirically revisable sentences ("Logic and experience"); towards the end of his life, it developed into the outlines of a new research programme in epistemology and in the philosophy of language ("The problem of empiricism and the concept of meaning").

So far as epistemology is concerned, Ajdukiewicz's philosophy during the second period of his philosophical life seemed to waver between moderate conventionalism and serious entertainment of radical empiricism, to progress finally towards a pluralist epistemology, i.e. the view that the duty of an epistemologist is not to take and defend one of the positions in the radical apriorism–radical empiricism spectrum (see his *Problems and Theories of Philosophy*) but rather to elaborate and thereby understand many possible conceptions of language and knowledge which could be used either to account for past and existing science or to anticipate further developments within its epistemological and linguistic framework.

So far as language and its philosophy are concerned, during the post-radical conventionalism period Ajdukiewicz made more extensive use of the ideas developed within logical syntax and logical semantics, though the pragmatic aspects of language continued to be at the centre of his interests ("Axiomatic systems from the methodological point of view", *Pragmatic Logic*). As an indication of Ajdukiewicz's gradual withdrawal from radical conventionalism one may perhaps mention the fact that in 1937 ("A semantical version of the problem of transcendental idealism") and again in 1948 ("Epistemology and semiotics") he gave semantical paraphrases of the doctrines of subjective and logical idealism and a critique of those doctrines based on his translation, though he was careful to admit that he could not be certain of the adequacy of the translation and that the authors of those doctrines would in all probability not regard his translations as adequate. In the rest of this essay we shall have the opportunity to comment on those aspects of Ajdukiewicz's conception of language which were fundamental to radical conventionalism and on those which he was able to retain in the post-radical conventionalism period.

From the point of view of what is currently fashionable in contemporary philosophy of science, radical conventionalism may seem more congenial than Ajdukiewicz's mature philosophy of the later period. At any rate it should be interesting to know how radical conventionalism originated, why it seemed at first so attractive to its author and what reasons or arguments persuaded him later to modify his philosophy. Moreover, because of the importance of conventionalist philosophy both for the development of science and of the philosophy of science, it is instructive to see how radical conventionalism compares with the philosophy of French conventionalism (commodism) and what was implied by either. In the rest of the present essay I shall outline first those philosophical views of Henri Poincaré and of Edouard LeRoy which were relevant to radical conventionalism, then discuss in more detail radical conventionalism and finally the transition to the post-radical conventionalism of Ajdukiewicz.

2. The Conventionalism of Henri Poincaré

By choosing the name of radical conventionalism for his earlier philosophy and by making explicit references to Poincaré's philosophy and to his polemic with LeRoy, Ajdukiewicz left no room for doubt that the views on mathematics, science and language of those two French philosophers were in some respects similar to his own.

Poincaré's whole philosophy was rather complex. He was a Kantian in his views on the epistemological status of arithmetic, claiming that some of the axioms of arithmetic, in particular the principle of mathematical induction, were synthetic *a priori* truths. On the other hand, he rejected Kantianism in the philosophy of space, geometry and physics, replacing it by a combination of genetic empiricism (concepts and statements in geometry and physics originated from experience) and conventionalism. In the foundations of set-theory his position was anti-Cantorian, constructivist, pre-intuitionist. In the philosophy of physics his conventionalism left room for empirical elements and so was within the bounds of the empiricist tradition. It was also coloured by many neo-Kantian and evolutionary ideas, like most philosophical doctrines of the time.

It is mainly Poincaré's philosophy of geometry and, to some extent, of physics that is of relevance here.

In his philosophy of geometry Poincaré argued both against Kantianism and empiricism. In conclusion he arrived at the following doctrines:

1. The axioms of Euclid's geometry, though they originate from empirical generalisations, form an implicit definition of the primitive terms of the system (e.g. "point", "lies between", "is equidistant"); they are *terminological conventions* neither true nor false but more or less convenient (*commodes*) (1902, 1952: p. 50); the same applies to the status of the axioms of other geometries.[2]

2. Alternative systems of metric geometry are different metric systems or metric languages which are *translatable* one into another on the basis of a suitable dictionary (1902, 1952: pp. 42–3).

3. Mathematical space to which we refer physical phenomena in our physical theories is a mathematical continuum (an idealisation of the "physical continuum" of our sensations) in itself amorphous; it can be metrised only if specific conventions are laid down concerning "congruence" or "distance"; this can be done in different ways yielding either Euclidean or non-Euclidean geometry—hence the

conventionality of metrics and of metric geometries (1902, Chap. II; *The Value of Science*, 1905, 1958: Chap. III, p. 37).

4. From the group-theoretic point of view one can say that geometries (metric and non-metric) are studies of invariants under various groups of transformations.[3] With respect to metric geometries it was pointed out by S. Lie that the congruence of two figures means that they are able to be transformed the one into the other by a certain point transformation in space; moreover, the properties in virtue of which congruence is an equality depend on the fact that displacements of figures are given by a group of transformations.[4]

5. What is *a priori* is the general concept of group; however, it is not an *a priori* form of sensibility but rather of understanding (in Kant's sense); within the general concept of group we can choose one particular group of transformations which will determine our geometry.

Poincaré used exactly analogous arguments against the Kantian and the empiricist views of the laws of mechanics and concluded that their status as well as that of "absolute time", "absolute space", was conventional (1902; 1952: Part III, Chap. 6; "De la mesure du temps", *Revue de Métaphysique et de Morale*, 1898, 1905). Moreover, he extended the group-theoretic viewpoint to physics. The principle of relativity, according to which the laws of physical phenomena are the same for a fixed observer as for an observer who has a uniform motion of translation relative to him (so that we have not nor can we possibly have any means of discovering whether or not we are carried along in such a motion), is equivalent to the principle that all laws of physics should be invariant under Lorentz transformations; in effect, physics is a study of invariants of the Lorentz group (which leaves invariant the form $x^2 + y^2 + z^2 - c^2 t^2$). The preference for the Lorentz group over the Galilean transformations, under which the laws of classical dynamics remain invariant, is not merely a matter of experimental findings but also of simplicity: Maxwell's equations are not invariant under the latter but are invariant under the former and the laws of mechanics are given a Lorentz invariant form in a simple way.

Apart from the discovery of non-Euclidean geometries, it seems that there were two other important sources of conventionalism in the philosophy of physics of Poincaré (and Pierre Duhem): one was an essentially *neo-Kantian conception of (empirical) meaning* (shared, however, both by pragmatists and by many nineteenth- and twentieth-century empiricists); the other was their reading of the history of science coupled with other arguments to the effect that there have been and always will be both *observationally equivalent and experimentally indistinguishable theories* in science.

The mentioned conception of (empirical) meaning may be stated, for example, in the following way. If two sets of sentences, S_1 and S_2, are observationally equivalent (i.e. have identical observational consequence classes), then they have the same scientific meaning or content; or—in a stronger form—the problem of choosing between two theories T_1 and T_2 is empirical if and only if T_1 and T_2 are neither observationally equivalent nor experimentally indistinguishable, given available observation and measurement techniques.[5] (See, however, the next paragraph.)

The relativity of motion, and the associated principles of relativity, starting with the kinematic or visual principle of relativity, known already to ancient Greeks, and including the more recent ones (e.g. the one mentioned before, the Poincaré principle) were referred to in support of the claim of the existence of observationally equivalent theories. But any, apparently rival theories, pretending to account for the *nature* of matter, light, etc., were regarded by both these authors as possibly observationally equivalent. Now, *observationally equivalent*—and even experimentally indistinguishable—*theories* according to Poincaré differ one from another in their theoretical part only linguistically, they are different *façons de parler*, they are alternative *languages*, more or less convenient, more or less inspiring or misleading (since they contain fictions and metaphors); again, as in the case of geometries, the question of truth or of falsity does not arise. The only way *a language* in a sense reflects reality is through its structure; in the transition from an old theory to a new one metaphors may change while the empirical laws as well as the relations expressed by differential equations of the theory's abstract part may remain true (1902, 1952: p. 161). In this sense the growth of science *is* cumulative rather than disruptive despite the ephemeral nature of theories (1902, 1952: pp. 160–5). Physical reality—the world—is knowable only up to the observational equivalence of alternative theories and up to the isomorphism of their theoretical postulates. This is essentially the Kantian sceptical doctrine concerning the limits of knowledge, expressed in a new form.[6] The term "instrumentalism", sometimes applied to this view of scientific theories (e.g. by Karl Popper[7]), may be misleading since it overemphasized the pragmatist element explicitly renounced in Poincaré's polemic with LeRoy and in his praise of the ideal of "pure science" (1905, 1958: last Chap.).

Poincaré did not classify either the axioms of geometries or the conventional principles of science (empirical laws elevated to the status of conventions) as analytic, presumably because traditionally analytic sentences were regarded as analytic truths whereas Poincaré denied any truth-value to both implicit definitions and to principles. However, Poincaré's philosophy had a tremendous impact on the later development of the idea of analyticity: firstly, when the concept of analyticity was suitably extended to cover all terminological conventions (Ajdukiewicz sentences dictated by axiomatic meaning rules; Carnap's meaning-postulates), analytic sentences came to be seen as determinants of language; secondly, Poincaré's genetic empiricism combined with conventionalism and his insistence that statements in science have been changing their status (perhaps the most famous case in point being the principle of relativity which—Poincaré claimed—had been an empirical generalization elevated to the status of a convention and then was threatened in this status by the outcome of Kaufmann's experiment[8]) made it necessary to relativise analyticity to language understood in a more precise and rigid way (Ajdukiewicz, Carnap) or else to question our ability to distinguish between the analytic and non-analytic sentences (Quine, White and others[9]) if "language" is used in the usual loose and amorphous sense; finally, Poincaré was one of the first, if not the first, to suggest that conventional and non-conventional components occur in one and the same sentence and might be split artificially (1905, 1958: Part III, Chap. 10, p. 124).

Ajdukiewicz's interest in Poincaré's philosophy antedated by at least fifteen years his articles on radical conventionalism published in the nineteen-thirties. In several papers read at the meetings of the Philosophical Society in Lwów in 1919 and 1920, as well as in his habilitation dissertation entitled *Z metodologii nauk dedukcyjnych* (Lwów, 1921; English translation "From the methodology of deductive sciences", *Studia Logica* **XIX** (1966)) Ajdukiewicz discussed various problems in the foundation of mathematics and logic and tried to clarify various foundational concepts (e.g. the concept of proof, entailment, mathematical domain, satisfaction, truth, existence, etc.). In the last of the three essays of which that dissertation consists, viz. in the essay entitled "On the Notion of Existence in the Deductive Sciences", he criticised Poincaré's claim, expressed in *Science and Method* (1908, 1956) and in three articles in *Revue de Métaphysique et de Morale* (**XIII, XVI**), that "existence in deductive theories amounts to consistency" (in mathematics "exists" means "is free from contradiction"). Against Poincaré, Ajdukiewicz argued that though consistency is a necessary condition for mathematical existence it is not a sufficient condition; "exists" in mathematics means "is an element of a proper (intended) domain of the specific mathematical theory". Ajdukiewicz's critique seems interesting for at least two reasons: firstly, his explication of "exists" with respect to mathematical theories is, in a sense, more constructivist and relies more heavily on an idea of mathematical intuition than Poincaré's purely formal requirement of consistency. Secondly, by insisting that absolute existential claims in mathematics make no sense, only relative ones do ("... in the deductive sciences we do not speak of existence in absolute sense but only relatively to a given system", "We may speak only of existence in a system as we speak of inclusion in a domain", "... there exist Euclidean straight lines and *non*-Euclidean straight lines; however, both cannot co-exist", "... the role of an existential postulate ... consists in enumerating those objects ... which may be substituted for the variables in the axioms", p. 45), he not only anticipated the thesis of "ontological relativity", at least with respect to mathematics, but also left a clue as to how a critical analysis of Poincaré's views in the foundations of mathematics and science may have inspired his linguistic relativism or conceptual perspectivism essential to radical conventionalism.

First of all, Ajdukiewicz obviously accepted Poincaré's general conventionalist thesis according to which there are conventional elements in our knowledge and there are problems not solvable without such conventions. However, he radicalised this claim, as we shall see later. He also agreed with Poincaré that the conventional elements in our knowledge are not isolated conventions but rather close-knit conceptual systems of language, which play a fundamental inferential role apart from being able to describe relations between the phenomena and, possibly, directly unobservable structures. It seems that Ajdukiewicz was influenced by Poincaré's preoccupation with the changes in science and mathematics, though he did not accept Poincaré's view of the fundamentally cumulative, non-disruptive nature of such changes. This is because Ajdukiewicz rejected both Poincaré's conception of meaning (of theories) and of language. Poincaré's discussion of alternative metric geometries as intertranslatable languages and his imaginary accounts of the communication between the inhabitants of non-Euclidean worlds must have been heuristically important. But the idea of the relativity of mathematical existence (in hyperbolic geometry rectangles do not exist, for example, and one cannot speak about rectangles in the language of that geometry), coupled later with an entirely different concept of language and of meaning (the origins of which may be found in Ajdukiewicz's 1921 dissertation), presumably persuaded Ajdukiewicz to disagree with Poincaré's and to agree with LeRoy's view on the role of the change in linguistic conventions and on the question of the existence of languages which are not mutually translatable. An outline of the dispute over this last problem will be given presently in the section dealing with LeRoy's conventionalism.

3. LeRoy's Nominalism and the Controversy over the existence of a "Universal Invariant"

Edouard LeRoy,[10] mathematician and philosopher, combined the most extreme elements of conventionalism of the new critique of science with Bergsonian evolutionism and life-philosophy in order to argue for a spiritualist "philosophy of freedom", for the superiority of intuition and religious experiences over science. It is his extreme form of conventionalism as well as some of his views on language that are of relevance here.

In "Science et philosophie" (*Revue de Métaphysique et de Morale*, **VII**, (1899)) and in "Un positivisme nouveau" (*Revue de Métaphysique et de Morale*, **IX**, (1901)), LeRoy presented a very extreme version of a conventionalist philosophy of mathematics and of science as the result of "the new critique of science" referring indiscriminately to the works of E. Boutroux, Poincaré, Duhem, G. Milhaud, J. Wilbois as well as his own.[11] Poincaré felt it necessary to reject—in a polemical article—LeRoy's "nominalism" and to clarify by contrast his own position ("Sur la valeur objective des théories physiques", *Revue de Métaphysique et de Morale*, (1902); *The Value of Science*, 1905; Part III, "The objective value of science"; the September issue of *Revue de Métaphysique et de Morale*, (1900) is likewise relevant). Pierre Duhem claimed priority over the other authors mentioned by LeRoy, though he carefully disassociated himself from their extra-scientific conclusions and suggested that Poincaré's reply was intended as much for him as for LeRoy.[12]

The philosophy of science reported with approval by LeRoy as the result of the new critique of science consists of: (1) the critique of "facts"; (2) the critique of "laws"; and (3) the critique of "theories".

1. *The Critique of "Facts"*

The traditional empiricist and Comtian positivist philosophy of science, perpetuated by numerous textbooks on the subject, contrasted "positive science" consisting of facts, with theories and hypotheses. Facts were claimed to be objective, theory-independent and simply discovered, stated and collected by impartial observers. However, Claude Bernard already pointed out that this is a naïve view of science since abstract concepts and theories play an important role in the study of even the simplest facts (LeRoy, 1899: p. 513); constant preoccupation of philosophers and scientists with rigour and objectivity too often results in the failure to see the role of the *free activity of the mind* in the determination of experimental truth (LeRoy, 1899: Part II, p. 514). Hence the need for a *new theory of scientific truth*. A visit to a scientific laboratory would show that scientists themselves have the impression of *constituting* facts, of *creating* from some amorphous material the particular objects of observation (LeRoy, 1899: Part II, p. 515; 1901: p. 145).

Reality is not accessible directly to the observer but only through the mediation of conceptual forms or schemes which are contingent on our past experiences as individuals and as a race, on our aims and prejudices as men of action, on everything life has imprinted on our minds. There are no isolated objects in nature, everything is diffused in everything. The process of separation or of objectification of nature, necessary and convenient for our thinking expresses ultimately nothing but the weakness of our bodies, exigencies of our actions or the variety of our points of view (LeRoy, 1899: Part II, p. 516). Our practically oriented senses are filters which leave only a residue of reality in our experience. There are no absolute facts, intrinsically definable as such. Any result of isolation, classification or approximation is relative to a viewpoint chosen in advance. In conclusion, far from being received passively by our mind, *facts are* in a way *created* by it (LeRoy, 1899: Part II, p. 517).

There is a difference, however, between commonsense facts and scientific facts; the latter are *theoretical interpretations* and differ from the former in the degree of conventionality, artificiality and relativity. The sentence "All men are mortal" may serve as an example of the former; "Phosphorus melts at 44°C", "The earth rotates" are examples of the latter. A "bare" or "pure" fact does not become a scientific fact unless it is placed within a system of concepts (LeRoy, 1901: p. 145). Admittedly, there is a mysterious residue of objectivity in facts, but science is not concerned with facts from the point of view of objectivity but rather as artefacts, as "atoms" in conventional schemes representing various points of view on nature (LeRoy, 1899, Part II: p. 518).

2. *The Critique of "Laws"*

According to the naïve traditional view, laws exist objectively, imprinted on facts; all that is necessary is to discover them among many irrelevant details.

In fact, laws are constant *relations*, or—to use a mathematical metaphor—they are *invariants* under universal transformation; they represent the element of stability in our changing experiences. Logically, they are abstract formulae, or schemes of classification, which summarise in a convenient way individual occurrences; they are also *mnemonic instruments, shorthands.* Furthermore, laws are second-order facts—more simplified, abstract and so more convenient; they replace first-order facts thus removing us further from the immediate contact with nature. All laws, far from being simply abstracted from things, are constructions of our mind, symbols of our ability to vary infinitely viewpoints from which the constancy of the world may be seen; once laws have been formulated, the situation is reversed: now facts are nothing but intersections of laws (LeRoy, 1899, Part II: p. 520). Are laws objective?—They are above all the results of our discursive operations which they express more than nature (LeRoy, 1899, Part II: p. 523). Laws as we know them are *contingent on* general forms (concepts) of commonsense such as *space, time, motion* whose sense we understand relative to practical needs of discursive thinking and of action, individual or social; they also depend on other conditions such as the *concepts of Euclidean geometry* which modern research has shown to have originated from the peculiarities of our experiences and from the habits of our life. Moreover, in so far as laws are quantitative they express the *conventions* on which measurements are based. Methodologically speaking, laws are *neither verifiable nor falsifiable*; they have often been saved from falsification by inventing new entities or by pleading that the conditions have not been exactly satisfied for the occurrence of predicted phenomena, or finally by pleading the limited sensitivity of our observational and measuring instruments. The physicist wants to have regularities, this is why he finds them with the help of ingenious violence to which he subjects nature (LeRoy, 1899, Part II: p. 523; 1901: p. 143). *Laws are implicit definitions of terms* (LeRoy, 1901: p. 143).

Are then scientific laws nothing but pure games or caprice of our mind? Such a claim would be untenable; they are, at least, *expressions of organised life* and of an *intentional attitude of our mind* (LeRoy, 1899, Part II: p. 525).

3. The Critique of "Theories"

Theories are *schematic and symbolic representations of* laws, *implicit definitions*; they are unverifiable and not subject to empirical control; they are rules of grammar; rival theories are like different dialects (LeRoy, 1899, Part II: p. 528).

From any theory by small variations one may construct infinitely many *observationally equivalent* or *experimentally indistinguishable theories* (LeRoy, 1899, Part II: p. 529).

Rival theories are "innumerable solutions of the same indeterminate problem" (Milhaud), they are distinct but *intertranslatable languages* (LeRoy, 1899, Part II: p. 529; 1901: p. 142) each has its advantages and scientists ought to be able to choose them according to their personal preferences; it is essential to have rival theories because each of them emphasises something that others disregard (LeRoy, 1899, Part II: p. 530). Theories, like laws, are neither verifiable nor falsifiable; once established, they become convenient languages (LeRoy, 1899, Part II: p. 531).

There is no true theory in the limit, only the preferred one (LeRoy, 1899, Part II: p. 532).

The aim of rational science is to construct a schema of the universe which would permit discursive thought to reproduce at will, without appeal to experience, all the developments of nature. Scientists strive to construct a table (diagram) with double entries in which symbols and phenomena correspond one to another, a dictionary of Real-Rational, like words of two languages coordinated for translation. The external world is conventionally represented, like a symphony by the score or celestial order by the differential equations (LeRoy, 1899, Part II: p. 544). The fundamental scientific ideas are those of *correspondence* and *representation*; two objects represent one another if any change in one is associated with a change in the other and, consequently, it is sufficient to act on one of them to induce an action on the other; two objects so related are logically substitutable, since knowledge of one suffices—by translation—to obtain knowledge of the other (LeRoy, 1899, Part II: p. 545).

In conclusion, the results of the new critique of science undermined the traditional belief in the necessity and objectivity of science. Facts, laws and theories are all conventional, relative, contingent. Those results have shown that scientific conclusions are not necessary unless one adopts a certain attitude; they explain the rigour of science by showing that the latter results from a decree of our mind, from conventions; freedom of our mind is the source of knowledge; in effect science is not shown to be purely arbitrary— except from a logical point of view—for science grows out of life, i.e. our practical and spiritual activities (LeRoy, 1901: p. 148). This is why *it is not possible to understand science as a purely intellectual enterprise*, for as such it is either merely a vast symbolism without any significance, or based on a vicious circle. One has to appeal to the *primacy of action, either practical or spiritual.*

The new critique of science has destroyed the old, naïve and narrow positivism of Comtian kind, which mistook commonsense, crude facts for immediate intuitions, and saw in industry, in production the only worthwhile activity. By its opponents the new critique was accused of *scepticism*. However, it has given birth to *New Positivism*, which is not sceptical: it is positivism in the sense that its highest ideal is also *action*, though not primarily practical action in the narrow sense but rather "profound action", i.e. the spiritual life (LeRoy, 1901: pp. 147, 151); the new positivism realizes that there are *diverse orders of knowledge* (the rational attitude is a legitimate but not the only and not the most valuable attitude one may take); even the most contrary views may contain correct elements; one must not reject outright a method merely because one has not been used to it; one can study things profitably from many different biases, there are many different ways of arriving at the truth. The superstition that there is one unique procedure, inevitable and perfect, usually results from lack of criticism (LeRoy, 1899, Part II: p. 377).

In his polemic with LeRoy Poincaré argued that: (a) as a philosophy of action LeRoy's extreme conventionalism or nominalism was self-defeating; (b) scientific facts are merely commonsense facts translated into a specialised, technical language of science and, therefore, not "created" by scientists in any clear sense; (c) there is a "universal, objective invariant" under changes of the conventions implicit in laws and theories, so that all conventional, theoretical languages are intertranslatable. ("Sur la valeur objective des théories physiques", *Revue de Métaphysique et de Morale*, 1902; *The Value of Science*, 1905, Part III, "The objective value of science".)

In connection with the first of these three points Poincaré simply argued that if it is true, as LeRoy following Bergson claimed, that science and its language have been geared primarily to practical actions, distorting reality in order to provide us with simplified recipes for actions, then it cannot be true that science consists exclusively of conventions. For conventions are arbitrary, free creations of our mind whereas practical recipes cannot be such. "There is no escape from this dilemma; either science does not enable us to foresee and then it is useless as rule of action; or else it enables us to foresee in a fashion more or less imperfect, and then it is not without value as means of knowledge" (1905, 1958: p. 115).

With regard to the second point, Poincaré's position was that the distinction between bare or rough (pure, commonsense) facts and scientific facts made by LeRoy (and Duhem) was not illegitimate. He objected, however, to LeRoy's (and Duhem's) view that rough or commonsense facts are outside science and of no relevance to it and that scientific facts, the only facts relevant to science were created by scientists in a conventional way, dependent merely on their choice of one of many possible theories (which are themselves nothing but conventions). In Poincaré's own view it is the rough or crude fact that imposes the scientific fact upon the scientist (1905, 1958: p. 116). The following are examples of statements of fact: (1) it grows dark (a report of a layman's impression of an eclipse of the sun); (2) the eclipse happened at nine o'clock; (3) the eclipse happened at the time deducible from the tables constructed according to Newton's laws. Of these, the first is supposed to represent a bare or crude fact, while the remaining ones are interpretations. To verify the first one, all we need is to specify the language in which it is formulated and to appeal to our senses, whereas in other cases (scientific facts or interpretations) we have also to specify certain additional *conventions*; but—so Poincaré argued—these conventions are merely rules of language, though more esoteric because used only by scientists and those who know science. In certain cases also some empirical laws are presupposed in the scientist's language which affirm, for example, that alternative methods of measuring (or detecting the presence of) certain effects will produce the same results: "To measure a current I may use a very great number of types of galvanometers or besides an electrodynamometer..."; should the underlying law (of the equivalence of results) turn out to be false "it would be necessary to change the scientific language to free it from a grave ambiguity". Poincaré concluded: "The scientific fact is only the crude fact translated into a convenient language" (1905, 1958: pp. 119, 120).

The last of the quoted claims was contested by Duhem on the ground that translation in this case would be indeterminate (to use a Quinean expression) and that the assumed laws which assure the equivalence of observational results form a theory, i.e. are conventional:

> It is therefore clear that the language in which a physicist expresses the results of his experiments is not a technical language similar to that employed in the diverse arts and trades. It resembles a technical language in that the initiated can translate it into facts, but differs in that a given sentence of a technical language expresses a specific operation performed on very specific objects whereas a sentence in the physicist's language may be translated into facts in an infinity of different ways." (*The Aim and Structure of Physical Theory*, p. 149)
> ...The role of the scientist is not limited to creating a clear and precise language in which to express concrete facts; rather, it is the case that the creation of this language presupposes the creation of a physical theory.... (*ibid.*, p. 151).

The distinction between "fact" and "interpretation" ("scientific facts") made by all three French philosophers was later to become the subject of Ajdukiewicz's criticism and the outcome of that criticism one of the bases of his radical conventionalism ("The world-picture and the conceptual apparatus", 1978: p. 67).

Finally, in his polemic with LeRoy, Poincaré credited him with formulating the problem of the *existence of "the universal invariant"* under changes of language, i.e. the problem of the *existence of languages* which are *not intertranslatable* (1905: pp. 126–7). "Since the enunciation of our laws may vary with the conventions that we adopt, since these conventions may modify even the natural relations of these laws, is there in the manifold of these laws something independent of these conventions and which may, so to speak, play the role of *universal invariant*?..." (1905, 1958: p. 127). In replying to this question Poincaré referred to fictitious non-Euclidean beings "educated in a world different from ours" and argued that their non-Euclidean language could always be translated into ours on the basis of a dictionary or—if they came to live with us in our world—both sides would end up understanding each other just "... as the American Indians ended by understanding the language of their conquerors after the arrival of the Spanish" (1905, 1958: p. 127). There were some conditions necessary for such understanding or translation, viz. some common humanity (e.g. similar sense organs), common principles of logic; there would be no translation, for example, "... if those beings rejected our logic and did not admit, for instance, the principle of contradiction" (1905, 1958: p. 128). Now, what is the nature of this invariant making all human-like languages intertranslatable?

"... The invariant laws are the relations between crude facts, while the relations between the "scientific facts" remain always dependent on certain conventions" (1905, 1958: p. 128).

As was pointed out in our account of LeRoy's philosophy, though he saw conventional (logically and empirically arbitrary) elements not only in laws and theories but also in scientific facts and was thus forced by the logic of his own thoroughgoing conventionalism to formulate the problem of translatability between alternative conventional languages, he explicitly admitted the intertranslatability (even exact translatability) of all theoretical languages without clarifying what the "universal invariant" under such translation would be; he did this possibly under the pressure of the authority of Poincaré's earlier writings on the subject. It was Duhem who questioned this universal translatability thesis having pointed out that a scientific fact (and interpretation) can be mapped onto infinitely many commonsense facts whose equivalence is always based on a suitable law, and a law was for Duhem part of a theory, i.e. of a conventional language. Thus by removing all bare or pure facts from science and by insisting that all scientific facts are theory-dependent (for their meaning and acceptance) both Duhem and LeRoy opted for a more subjectivist and relativist view of science which deprived the latter of its superiority—in terms of objectivity and rationality—over religious beliefs and other "spiritual experiences".[13] Poincaré—whose views are often misrepresented as thoroughly conventionalist[14]—remained faithful to the tradition of scientific objectivity admitting that one may adopt a strictly nominalistic attitude with respect to all theories, laws and facts but insisting at the same time that such a nominalist attitude is alien to science as we know it (1905, 1958: pp. 122–4). Conventions do play an important part in science; but once some conventions have been laid down, the problems scientists are interested in are of empirical nature and are solved on the basis of experimental procedures (1905, 1958: pp. 120, 122). Moreover, there is no sharp and unbridgeable frontier between science and common sense which we use in everyday life (1905, 1958: p. 122).

The controversy over the translatability of languages between Poincaré and LeRoy (involving also indirectly Duhem) must have had some influence on Ajdukiewicz's radical conventionalism. He himself refers to it (in "The world-picture and the conceptual apparatus", 1934) only in connection with the disagreement between Poincaré and LeRoy concerning so-called bare or pure facts, apparently siding with LeRoy's claim that scientists "create (scientific) facts", in fact arguing against all three French authors' distinction between bare (pure) facts (observational reports) and interpretations, and their implicit belief that one might be free to change more artificial scientific conventions but not those traditionally embedded in everyday language.

It is plausible to suggest that LeRoy's challenge to Poincaré's thesis of the translatability of all human languages and his Bergsonian and Pragmatist conception of language as "part of life", as expressing various "attitudes, ways of life, points of view" and associated with "action", did exert some influence on radical conventionalism. If so, then one must also emphasize the counterbalancing influence of the mathematical logicians's conception of language as uniquely given by its vocabulary and rules. While Poincaré and LeRoy discussed translatability (or otherwise) of languages in terms, firstly, of "shared humanity" of speakers (e.g. having sense-organs like ours), secondly, of the properties of the worlds in which they lived (Euclidean, non-Euclidean worlds), and, thirdly, of the existence of bare or pure facts, Ajdukiewicz based his thesis of the existence of not-intertranslatable languages on more rigorous conceptions of language, meaning and translation. Finally, notwithstanding the latter thesis, science according to Ajdukiewicz's radical conventionalism is far *less* arbitrary and subjective than LeRoy would have had it.

4. The Ajdukiewicz Languages and Radical Conventionalism

According to Ajdukiewicz's own statement (in "Language and meaning", and "The world-picture and the conceptual apparatus"), the epistemological theses of radical conventionalism are consequences of the definition of "meaning" associated with the conception of language which he developed in "On the meaning of expressions" and in "Language and meaning". Different conceptions of meaning give rise to different views concerning the cognitive role of language. Some of them imply that the cognitive role of language reduces simply to its role as a means of communication and since all sufficiently rich languages are mutually translatable, the choice of one of them does not affect our knowledge in any essential way, though it may make a difference in other respects. Such, as we have seen, was the view of Poincaré. But there is another view, according to which the cognitive function of language is of crucial importance and transcends that of a means of communication being comparable to the role Immanuel Kant attributed to the *a priori* forms of sensibility and of understanding (Ajdukiewicz, 1978: p. 86), except that there are, so it is claimed, *many different, mutually not translatable languages* which may provide completely *different world-perspectives* when combined with suitable experiential data. Such different linguistic world-perspectives might perhaps be compared to the radically different world-pictures envisaged briefly and dismissed by Poincaré (in his polemic with LeRoy) when he mentioned beings with sense organs different from ours (more sensitive to some, less sensitive to other stimuli) and using a logic different from ours (Poincaré, 1905, 1958: III, Chap. X, p. 128). This other view, according to which languages play a much more dramatic role in cognition, led Ajdukiewicz to the doctrine of radical conventionalism.

Language as actually spoken by men is for Ajdukiewicz—as it was for LeRoy—part of our biological and social life, adaptive behaviour, goal-directed and rule-governed activity. Apart from the rules of phonetics and syntax, the language of a community of speakers might be seen as governed by the *rules* which would express the speakers' *dispositions* to definite *motivational relationships* whereby experiences of certain types *motivate the acceptance* of definite sentences of the language (this is a pragmatic conception of inference). However, an empirical study of language—wherein to discover if a speaker associates with sentence S the meaning assigned to it by the meaning rules of L one places him in a situation appropriate to S and elicits his assent to or dissent from S (Ajdukiewicz, 1978: p. 40)—would soon reveal that in a natural language there are no fixed and universal motivational relationships, in other words, that a natural language should rather be seen as a family of languages each of them with fixed and binding rules, each of them corresponding perhaps to a different way or form of life (languages of different social strata, professions, etc.). However, there is more to language than the psychological processes. The logical aspect (judgements in the logical sense), relevant to knowledge in the objective sense, rests in the intersubjective meaning which expressions have relative to a language-structure and not relative to individual speakers. If this is granted, then a transition is made to a more abstract study of language in which the latter is an idealisation ("idealised type"), intended to reflect the inherent, though latent structure of a part of a natural language (Ajdukiewicz, 1978: pp. 26, 63). In an abstract study of this type, language is reconstructed in terms of its vocabulary, the rules of syntax and the *meaning-specification*, or the *meaning*-(acceptance) *rules*. The latter determine the *structure* or *matrix of the language* and the *meaning* of expressions is then conceived as an abstract property they have in virtue of the *positions* they may take *in the matrix* (Ajdukiewicz, 1978: p. 62). Three kinds of meaning-rules may be distinguished as basic. (1) *Axiomatic meaning-rules* specify sentences which are to be accepted unconditionally: the rejection of any sentence dictated by an axiomatic meaning-rule amounts to the violation of the meaning-specification characteristic for the language. (2) *Deductive meaning-rules* specify ordered pairs of sentences (or ordered pairs whose first element is a sentence-class and the second element is a sentence) such that if one accepts the first of them one is thereby committed to accepting the second on pain of violating the meaning-specification of the language. (3) *Empirical meaning-rules* assign to definite experiential data sentences (simple empirical meaning-rules) or to definite experiental data and sentences, other sentences (compound empirical meaning-rules) such that in the presence of those data (possibly conjoined with the acceptance of some sentences), one is forced to accept the coordinated sentence if one is to avoid violation of meaning. Axiomatic and deductive meaning-rules are *discursive rules*, they are sufficient for purely discursive languages, e.g. of pure mathematics (Ajdukiewicz, 1978: pp. 47, 68). Each meaning-rule has a *scope*: the scope of an axiomatic meaning-rule is a set of sentences (the axioms or principles of the language); the scope of a deductive meaning-rule is a set of ordered pairs of sentences (or of sentence-class/sentence); the scope of an empirical meaning-rule is a set of ordered pairs of experiential data and sentences (or experiential data-*cum*-sentences-sentence). The scopes of the meaning-rules of the same type may be summed. The sum of the scopes of all meaning-rules of the same type is their *total scope*. Two expressions of a language have *the same meaning* (are synonymous or intertranslatable) if their interchange leaves the total scopes of the meaning-rules unaltered, i.e. if they are *isotopes* in the total scopes of the meaning-rules. This statement may be made more precise in terms of the matrix of the language (Ajdukiewicz, 1978: p. 62). The meaning-specification of a language is the correlation between its expressions and their positions within the scopes of the meaning-rules (and, consequently, within the matrix). Thus for a change of the meaning-specification of a language a change in the total scope of at least one of the types of meaning-rules is necessary. Two expressions of a language are said to be *directly* (*immediately*) *meaning-related* if they both occur in the same element of the scope of a meaning-rule; they are *indirectly* (*mediately*) *meaning-related* if they are linked by a finite chain of expressions which are directly meaning-related among themselves. A language in which any two expressions are directly or indirectly meaning-related is called *connected*; a *disconnected* language is an assemblage of connected languages. A language L is said to be *open with respect* to another language L' if L' contains all the expressions of L with the same meanings they have in L and also some expressions which do not exist in L and such that at least one of them is directly meaning-related to an expression in L. A language is *open* if there exists another language with respect to which it is open. A language which is not open is *closed* (Ajdukiewicz, 1978: p. 50). The distinctions just introduced are crucial for Ajdukiewicz's analysis of *meaning-change* ("variance") and thus of *language-change*, as well as for his analysis of the relations between different languages. In effect, they are crucial for the theses of radical conventionalism.

Not every change in the scope of individual meaning-rules results in the change of the total scope of the meaning-rules: different individual scopes may yield identical scope-sums; hence there are changes in the individual scopes which leave the meaning-specification of the language (and thus the meanings of its expressions) unaltered (Ajdukiewicz, 1978: p. 49). When the meaning-specification of a language does change, this can occur either through the introduction of new expressions or without such enrichment of the vocabulary. The latter change occurs, for example, when a sentence which originally was an empirical hypothesis is elevated to the status of a principle (convention) in Poincaré's sense. According to Poincaré, we may remember, the postulates of Euclid's geometry, the laws of Newtonian mechanics, conservation laws, the principle of relativity all changed their status in this way at one time or another. However, from Poincaré's point of view all these changes occurred within one and the same language, or at least did not involve any change in the meaning of the sentences in question. By contrast on Ajdukiewicz's conception of language and meaning a radical change in meaning occurs whenever the status of a sentence is altered in this way; for the rejection of an empirical hypothesis does not violate any meaning-rule whereas the rejection of a sentence dictated by an axiomatic meaning-rule, e.g. a principle or terminological convention, does. According to Ajdukiewicz's view of language it is logically impossible to reject a sentence dictated by an axiomatic meaning-rule, i.e. a sentence analytic in the given language, if one understands it in conformity with the meaning-specification of the language. For a proper use of a sentence dictated by an axiomatic meaning-rule requires that it should be accepted (Ajdukiewicz, 1978: p. 111).

As regards the introduction of new expressions into a language's vocabulary, an open language behaves under such a change differently from a closed one: one can introduce into an open language L a new expression E which is not synonymous with any expression already in L and such that E is directly meaning-related to some expression L, without thereby altering the meaning of the old expressions of L. A closed language, on the other hand, becomes disconnected if it is enriched with a new expression synonymous with some of those already in the language (Ajdukiewicz, 1978: p. 51).

If two languages L and L' are both closed and connected and if some expression of L has a translation in L' then the two languages L and L' are *intertranslatable*. An open language can close to two closed and connected languages which are mutually translatable but cannot close to two closed and connected languages which are not intertranslatable. If the totality of the expressions of a closed and connected language together with their meanings (i.e. positions in the total scopes of the meaning-rules) is called a *conceptual apparatus* and the totality of accepted sentences of such a language a *world-picture*, then the *theses of radical conventionalism* may be formulated as follows:

> Of all the judgements [meanings of sentences—J.G.] which we accept and which accordingly constitute our entire world-picture, *none* is unambiguously determined by experiential data; every one of them depends on the conceptual apparatus we choose to use in representing experiential data. We can choose, however, one or another conceptual apparatus which will affect our whole world-picture (Ajdukiewicz, 1978: p. 67; for a more precise statement, see same article, Chap. 3, p. 72).

Apart from the above doctrine of radical conventionalism, Ajdukiewicz also subscribed to a weaker thesis of radical conventionalism, formulated in his polemic with Poincaré and other French conventionalists, and independent of the concept of closed and connected languages which is essential for the formulation of the stronger theses of radical conventionalism. The weaker thesis of radical conventionalism is as follows:

There is no essential difference between so-called observational reports (bare or pure facts) and interpretations so far as their epistemological status is concerned; the former are governed by the meaning-rules of the ordinary language, whereas the latter are governed in addition by explicitly introduced conventions whose purpose is to make the expression of the ordinary language more precise (Ajdukiewicz, 1965: "W sprawie artykulu prof. A. Schaffa o moich pogladach filozoficznych", *Jezyk i Poznanie*, Vol. II, p. 181).

If we call *universal* a language in which all judgements (meanings) can be expressed, then science in its actual development does not show a tendency towards universality; a universal language would be disconnected, a loose assemblage of conceptual apparatuses. Science tends towards a connected world-view (Ajdukiewicz, 1978: p. 79).

A conceptual apparatus together with a set of experiential data is called *a world-perspective* (corresponding to those data), (Ajdukiewicz, 1978: p. 112). The set of sentences dictated by the meaning-rules of a conceptual apparatus (together with well-confirmed hypotheses framed in the concepts of the language) form *a world picture* (Ajdukiewicz, 1978: p. 81). Two different world-pictures (associated with conceptual apparatuses which are not mutually translatable) are neither logically nor experimentally comparable. One could extend a conceptual apparatus L by introducing a *concept of truth*, perhaps with the help of the following meaning-rule: Only that person speaks the language L who having accepted a sentence S of L is on that basis prepared to accept the sentence "S is true in L". So introduced, the concept of truth would apply to the sentences of L only. Using one conceptual apparatus one cannot predicate "truth" or "falsity" of any sentence belonging to the world-picture associated with a different conceptual apparatus, not translatable with one's own (Ajdukiewicz, 1978: p. 117).

If different world-pictures cannot be compared either logically (through inference or with respect to their truth or falsity) or experimentally, are they equally good or can they not be compared in any way whatever?—They can be compared and evaluated in the process of "humanistic understanding" (Ajdukiewicz, 1978: p. 117) or from an "evolutionary" point of view. The former is in effect a study of the motives which guide scientists in bringing about changes in science. The latter is concerned with the question: Which of the conceptual apparatuses is closer to the (*quasi-*)goal discernible from the evolutionary tendencies in actual science? (Ajdukiewicz, 1978: p. 117).

Let us label as *Ajdukiewicz languages* the (idealised) languages given by the vocabulary, the syntax and the three types of meaning-rules, as described before.[15] The Ajdukiewicz languages have several interesting properties, among them the following ones: (1) they are axiomatic systems; (2) for any of the language speakers they are axiomatic systems not only in the formal but also in the pragmatic sense; (3) assuming Ajdukiewicz's definition of meaning, they are interpreted axiomatic systems in a sense of 'interpreted' different from the usual semantical-extensional one; (4) they are theories in a literal and quite precise sense.

The Ajdukiewicz languages are *axiomatic systems* since the meaning-rules of each such language define a relation of consequence on the class of its sentences. This is obvious in the case of languages which are based only on discursive (axiomatic and deductive) meaning-rules. The empirical meaning-rules, however, can be reduced to axiomatic ones, as has been shown by Ajdukiewicz (in 1936: Empiryczny fundament poznania"; 1953: "W sprawie artykulu prof. A. Schaffa o moich pogladach filozoficznych", *Jezyk i poznanie*, Vol. II, p. 163).

Since the axioms of an Ajdukiewicz language have to be accepted by any speaker of the language together with all their consequences, an Ajdukiewicz language is for any of its speakers *an axiomatic system in the pragmatic* and not only in the formal *sense*.

In his 1921 dissertation Ajdukiewicz distinguished between the formal (syntactical), semantical and methodological (pragmatic) properties and studies of axiomatic systems, though at the time he did not use the terms "semantics" or "pragmatics". In a couple of articles published in the post-radical conventionalism period ("Axiomatic Systems from the Methodological Point of View", also 1948: "Metodologia i metanauka" translated as "Methodology and metascience" in 1976: *25 Years of Logical Methodology in Poland*, (edited by M. Przelecki and R. Wójcicki)), Ajdukiewicz discussed the differences between the meta-logical (formal and semantical) and the methodological aspects of axiomatic systems in mathematics and in science. From the methodological point of view one is interested in the pragmatic relations between the axiomatic systems and men who use them, e.g. in the question whether or not their axioms and derived theorems are asserted (accepted) by those who use them and how such acceptance is justified. Ajdukiewicz proposed to distinguish three types of axiomatic system from the methodological or pragmatic viewpoint: An axiomatic system is *assertive-deductive* (for X) if both its axioms and derived theorems are accepted (by X) and the acceptance of derived theorems is motivated by the acceptance of the axioms. An axiom system is *assertive-reductive* (for X) in case the acceptance of the axioms is motivated by the acceptance of some of the derived theorems (so-called hypothetico-deductive systems in science are of this type). Finally an axiomatic system is *neutral* (*non*-assertive (for X)) if neither its axioms nor its derived theorems are asserted (by X) but merely "assumed" or "entertained". Though this distinction was made by Ajdukiewicz not for semantical but for methodological purposes and no reference was made by him in that context to his earlier conception of language and "meaning" (understood in an inter-subjective sense), it is tempting to characterise the Ajdukiewicz languages in the light of this classification of axiomatic systems, and assuming Ajdukiewicz's definition of "meaning". On the assumption mentioned, the Ajdukiewicz languages are, of course, assertive-deductive axiomatic systems for any speaker of those languages (if one disregards semantic requirements, avoided in "Language and meaning"). However, since not all sentences of an Ajdukiewicz language are dictated by axiomatic meaning-rules— hypotheses, for example, are not—within a language of this type there may be embedded either an empirical theory or an abstract theory which is assertive-reductive or else neutral; this would allow one to treat not only empirical hypotheses, but also some mathematical axioms, e.g. the axiom of infinity, the axiom of choice, etc., as hypotheses rather than as sentences dictated by the axiomatic meaning-rules, i.e. as analytic.[16]

Assuming Ajdukiewicz's definition of "meaning", and disregarding for the time being certain inadequacies in it to be discussed later, an Ajdukiewicz language is an *interpreted system*, since every expression of such a language has meaning, i.e. position in the total scope of the meaning-rules which govern the acceptance of sentence. This is a sense of "interpretation" different from the usual semantical-extensional one, as a mapping of the expressions of the language onto an abstract or empirical domain (structure). For although every expression of an Ajdukiewicz language has meaning, not every one has a referent (denotation). Moreover, two co-extensive expressions, i.e. which have identical denotations in a domain, may have different meanings. Indeed, one of Ajdukiewicz's objectives was to do justice to these differences between "meaning" and "denotation" (1978: p. 39, also his reference to Mill's distinction between "denotation" and "connotation" there). As can be gathered from our previous comments, an axiomatic system may be interpreted in either an abstract domain, e.g. of numbers, or in an empirical domain, and yet remain uninterpreted in the present sense; this happens, of course, if it is a neutral axiomatic system with respect to whose axioms and derived theorems one takes no assertive attitude. Such a system, on its own, would *not* be an Ajdukiewicz language (in the present sense; see, however, Section V for the Ajdukiewicz languages without axiomatic meaning-rules), since its axioms are not dictated by axiomatic meaning-rules though it could be embedded in one to give its expressions meaning in Ajdukiewicz's sense. "Meaning" in Ajdukiewicz's sense was understood in a non-psychological sense of "objectified meaning", relativised only to a language. The latter in its logical aspect is a relatively autonomous entity. The actual linguistic performance of the language-users somehow participates in it, since the language-users are committed to the acceptance of those sentences of the language which its meaning-rules dictate. Linguistic expressions on this conception acquire meaning in a way analogous to the process in which "life-expressions" acquire "objective sense" (*Sinn*) within a cultural structure, according to the philosophy of the "objective mind". Meaning (interpretation) in this sense may be given directly to the most abstract expressions whether mathematical, scientific, theological or metaphysical in general. For what is essential to "meaning" so understood is not—as in the case of indirect interpretation through so-called "bridge-principles"—the relation of an expression to observables, but its position (valence) in a language-structure.[17] The methodological question of the justification of such commitments is distinct from the semantical one. Needless to say, the existence of referents is not guaranteed by having a meaning in this sense nor—as we shall see—is identity of referents of synonymous expressions. Finally and trivially—in virtue of previous properties—an Ajdukiewicz language is a *theory* in the clear and precise sense of a deductively systematised set of asserted sentences. Let us repeat, a user of an Ajdukiewicz language is committed to accept some of its sentences—those dictated by the meaning-rules—in order to understand its expressions in a way appropriate to the given language ("There is no understanding without some belief" as a contemporary hermeneuticist wrote). By contrast, Poincaré's view was that an abstract *theory* in theoretical physics is merely a *façon de parler*—a language of symbols and metaphors which *does not commit* one to any assertion, apart from the empirical laws invariant under change of theoretical language and possibly its invariant structure. Similarly, for Carnap, realism and idealism were two observationally equivalent metaphysical theories and thus two different *façons de parler* without any commitments. In at least one of the senses of "instrumentalism" a scientific theory from the instrumentalist viewpoint is a system of unasserted (though possibly semantically interpreted) sentences, i.e. it is a neutral (non-assertive) axiomatic system in Ajdukiewicz's terminology.[18] Such a view expresses either the instrumentalist's scepticism with respect to the specific scientific theory (or to any scientific theory) or else his commitment to an apparently incompatible metaphysical or religious doctrine and to the ideal of consistency of one's assertions. It is worth noting, first of all, that Poincaré's view of physical theories was not instrumentalist in this sense (whereas Bellarmino's and Osiander's, for example, were); secondly, that a sceptic-instrumentalist (in the sense just explicated) may use an Ajdukiewicz language to embed in it a neutral, unasserted scientific theory. Ajdukiewicz himself never held an instrumentalist view of scientific theories in any of the senses mentioned, except possibly the last one with respect to mathematics at the time when he wrote his "Axiomatic systems from the methodological point of view".

The stronger thesis of radical conventionalism implies that there are at least two different not intertranslatable conceptual apparatuses, i.e. closed and connected languages. Three concepts are important for the understanding and assessment of the truth of this claim: "meaning" (since "translation" is defined in terms of "sameness of meaning"), "closed language", "connected language". The first two turn out to be associated with certain difficulties which will be discussed in the final section of this essay. Ajdukiewicz's general conception of language as based on vocabulary, syntax and the meaning-rules (the Ajdukiewicz languages as we have called them here) was largely unaffected by these difficulties and continued to be used by him for various purposes in the later period of his life.

5. Beyond Radical Conventionalism

Radical conventionalism grew out of Ajdukiewicz's conviction that to solve epistemological problems in a satisfactory, clear and precise fashion one needs a *more rigorous conception of language and meaning* than those used in traditional epistemological discussions. For *knowledge* consists, in traditional terminology, of judgements and concepts and the latter are simply *meanings* which sentences and other expressions have relative not to individual knowing subjects but *relative to a language* (or a family of languages) (Ajdukiewicz, 1978: p. 35). The same applies to scientific laws and theories. Newton's law of gravitation, for example, is the meaning a particular sentence has in a specific language and which is shared by all its *translations*. The controversy between the apriorists and the empiricists as to whether there are *a priori* elements (judgements) in our knowledge and if so whether they are all analytic or not, concerns articulated knowledge and this presupposes a language, i.e. a system of expressions endowed with meaning (Ajdukiewicz, 1978: p. 39). One cannot hope to resolve the controversy without having a clearer notion of language and meaning. The same applies to a more specific controversy between the apriorists, radical empiricists and conventionalists (moderate empiricists) concerning the status of Euclid's axioms or of the laws of Newtonian mechanics. The controversy presupposes a concept of analyticity (as well as other related concepts) and, clearly, Kants's definitions were unsatisfactory, especially in the light of Poincaré's insight into the changes of the status of the components of mathematical and scientific theories (from empirical generalisations to conventional principles). This insight is wasted if one does not see those changes as indicative both of the changes in language and of the relativity of status. Finally, the dispute between Poincaré and LeRoy concerning the existence of the "universal invariant" under translation from one language into another and the question whether all languages are intertranslatable, is hopelessly undecidable as long as it is conducted in obscure and vague terms (e.g. "shared humanity", "universal invariant", and unanalysed concept of "translation"). However stimulating these controversies are, one should approach them in a more rigorous fashion if one takes them seriously.

In order to realise this research programme Ajdukiewicz set out in the early nineteen-thirties to construct a concept of language and of meaning at once more precise than the traditional ones and yet embodying some, at least, of the traditional insights. Language itself was to be seen as both rooted in psychological phenomena susceptible of empirical research (judging process, attitudes, assertive behaviour) and as transcending them into the sphere of intersubjective structures given by rules, viz. meaning-rules constituted by and abstracted from human activities. Meaning was conceived as an abstract, intersubjective property expressions have in virtue of their positions or role (inferential valence) in the language-structure. Language was thus both a humanistic or cultural structure knowable through participation in it (understanding, acceptance, assertion)[19] and a formal structure (an axiomatic system) which could be studied with the help of formal methods.

Having adopted this view of language in the first period of his philosophical development, Ajdukiewicz was in a position to argue that there were many linguistic structures or conceptual apparatuses and that one's world-perspective depended not only on experiential data but also on the choice of one such conceptual apparatus. Moreover, he was able to give a definition of analyticity in L (the set of sentences dictated by the axiomatic rules of L and their consequences by virtue of deductive rules) and to argue both that the alleged synthetic *a priori* sentences were analytic in his sense and that the presence of the *a priori* analytic sentences in our knowledge is unavoidable since it is simply due to the fact that knowledge has to be articulated in a language. In this way, *the clarification of the concepts of language and of meaning had important epistemological consequences*: the first was that one's world-view was co-determined by experiential data *and* the choice of a language; the second was that the solutions of various traditional epistemological problems were determined by the essential properties of language and were given once the relevant properties of language were exhibited in a satisfactory theory of language. The latter seemed also to give an extra bonus: semantical concepts such as "denotation" ("reference"), "truth", "falsity", etc. which had been known to generate paradoxes, were dispensable. "Meaning" was neither identified with nor defined in terms of "denotation". Individual sentences and world-pictures were not characterised as true or false but rather as dictated by meaning-rules, i.e. as derivable immediately or mediately by meaning-rules. If the term "true" were to be introduced at all, it would be governed by a meaning-rule which restricted its application to the sentences of one's language already dictated by the meaning-rules of the language (cf. Ajdukiewicz, 1960: Preface to *Język i Poznanie*).

Ajdukiewicz's articles on language, meaning and radical conventionalism appeared at the time when in Vienna logic as the method of philosophy was identified with logical syntax. However, their publication roughly coincided with the emergence of logical semantics: Alfred Tarski's memoir 'Pojecie prawdy w jezykach nauk dedukcyjnych' (*Towarzystwo Naukowe Warszawskie*, Warszawa) appeared in 1933 and soon, in 1936, was made more accessible in German translation as "Der Wahrheitsbegriff in formalisierten Sprachen" (*Studia Philosophica* I). Pragmatics, mainly empirical pragmatics and as a research programme rather than as an accomplished technique, was to be enshrined a bit later as a part of the logical empiricist theory of language (Ch. Morris, 1938: "Theory of signs").

The impact of logical semantics on Ajdukiewicz's philosophy was threefold. Firstly, he at once joined in the programme of developing epistemology based on logical semantics, since semantical concepts were now shown to be free from the paradoxes, provided the use of those concepts was suitably restricted. Secondly, he began to doubt whether a satisfactory definition of "meaning" could be given without the use of semantical concepts (in the narrower sense). Thirdly, he came to the conclusion that the classical conception of truth (explicated by Tarski) was essential for realist epistemology. As regards the first and third points, one should mention his article "A semantical version of the problem of transcendental idealism", (1937), in which Ajdukiewicz used both some results of semantics (e.g. "The class of the theorems of an incomplete theory is not identical with the class of true sentences in the language of the theory") and his own conception of language (meaning-rules as defining the relation of consequence on the set of sentences of a language) to paraphrase the thesis of transcendental idealism and to subject it to criticism. In the same article he observed that, if dissatisfied with the application of the law of the excluded middle to sentences with vague terms one were to replace "true sentence" with "provable theorem" (or "sentence derivable in virtue of meaning-rules"), then one would be committed to epistemological idealism (a similar claim may be found in *Problems and Theories of Philosophy* (1973) p. 21).[20]

Meanwhile it was pointed out by Tarski (as reported by Ajdukiewicz in "The problem of empiricism and the concept of meaning") that, at least for a language L based exclusively on axiomatic and deductive meaning-rules one can construct two expressions A and B such that the meaning-rules of L are invariant under the interchange of A and B and yet A and B have *non*-identical denotations. This shows that, if one demands (as one usually does) that two synonymous expressions should have identical denotations, then *the invariance of the meaning-rules under the interchange of expressions—though necessary—is not sufficient for those expressions of a language based on axiomatic and deductive meaning rules to be synonymous.* Let us remark, first of all, that the objection does not apply to languages without axiomatic meaning-rules. However, if one accepts Tarski's criticism—as Ajdukiewicz did—then "synonymity" and "meaning" are no longer explicitly defined in terms of the language matrix and the equivalence class of positions within the matrix. They are defined only *partially* by the implication: if two expressions of L are synonymous, then the meaning-rules of L are invariant under interchange of those two expressions. This implication, unaffected by the objection, was sufficient for Ajdukiewicz to be able to continue claiming that the presence in our knowledge of analytic (*a priori*) sentences is unavoidable since it is due to our use of language. The same implication also suffices as a criterion of meaning-change (variance), since it provides a criterion of the difference in meaning. Thus many problems of meaning-change discussed in contemporary philosophy of science under the heading of "incommensurability" can still be handled within the framework of Ajdukiewicz languages. Finally, the objection could have been met by making the relevant adequacy condition part of the definition of "synonymity": two expressions of L are synonymous if and only if the meaning-rules of L are invariant under the interchange of those expressions and if the denotations of the expressions in question are identical. In later years one of the alternative definitions of "meaning" which Ajdukiewicz considered, and which he thought would be neutral with respect to different epistemological positions, was in terms of "codenotation", i.e. in terms of the denotation of a compound expression and of the syntactical positions (within the expression) of its constituents. The latter concepts he developed in his very influential 1936 article "Syntactic connexion", and used in many later articles, including the posthumously published "Proposition as the connotation of a sentence" and "Intensional expressions".[21]

Apart from the mentioned objection to Ajdukiewicz's definition of "synonymity", some criticisms were also made of his concept of a closed language, i.e. of a language which cannot be further enriched without becoming disconnected. Critics argued that closed languages did not exist that—*a fortiori*—closed not-intertranslatable languages did not exist either. Since Ajdukiewicz's definition of "meaning" was given for closed and connected languages, this criticism may seem to be directed against the definition of "meaning" also. Now, the assumptions of connectedness and closedness were made by Ajdukiewicz with full awareness of their idealized nature (1978: pp. 64, 87). A disconnected language is not *one* language but a *family* of connected languages. Similarly, an open language is an "unfinished" language, a language which is in the process of growth or, at least, of potential growth; in other words, such a language is only part of a "complete language". It is, therefore, impossible to construct a matrix either for a disconnected or for an open language. In the former case it would have to be a set of matrices; in the latter, the matrix would not be complete and could be extended further in many different ways by introducing different meaning-rules for new expressions; consequently one could never know if the condition of the invariance of the total scope of rules of one type or another was satisfied or not.

In other words, the claim that closed languages did not exist could not be a serious objection to the idealised assumption of closedness whose sole function was to make sense of the concept of the language matrix (used in turn to define synonymity and meaning); all idealisations would be affected by similar objections. On the other hand, if substantiated, the claim would constitute a valid objection to one of the theses of radical conventionalism, viz. the thesis that there are at least two different conceptual apparatuses (i.e. connected, closed and mutually non-translatable languages). For though one may be permitted to make use of simplified assumptions in one's definitions and theories, one must not draw existential conclusions from such assumptions.[22]

These and similar criticisms naturally undermined Ajdukiewicz's confidence in radical conventionalism in its stronger version (the weaker was unaffected); they also made his views on language and meaning more flexible which in turn—not surprisingly—led him to embrace a position not unlike his original conventionalism except that it was now *conventionalism at the meta-level*. Radical conventionalism implied the existence of different mutually non-translatable object-languages, from which one at a time could be chosen to articulate our experiences and discursive thoughts. It originated from an analysis in which Ajdukiewicz believed to have captured, as it were, the nature of language in its cognitive function as it is known to us from everyday experience in practical life, in existing science and mathematics. At least three types of meaning-rule (axiomatic, deductive and empirical) are characteristic of any existing empirical language and two (axiomatic and deductive) of any known discursive language. But just as the concept of Euclidean space and geometry, though they originated from our everyday experiences (with rigid bodies), were supplemented by "non-Euclidean spaces" and *non*-Euclidean geometries, the standard concept of an empirical Ajdukiewicz language may be supplemented by the idea of a language based on deductive and empirical meaning-rules exclusively (Ajdukiewicz, 1947: "Logic and experience", p. 168).[23] In such a nonstandard empirical Ajdukiewicz language there are no analytic sentences since there are no axiomatic meaning-rules necessary to generate analyticity. The role of logic would be performed either by the deductive meaning-rules alone or also by formulae which would be used together with empirical premisses in inferences and which—in view of the Duhem thesis—would be affected by the outcome of the experimental testing of the conclusions. An Ajdukiewicz language without axiomatic meaning-rules is in effect a Quinean language for no sentence is accepted in it "come what may", i.e. unconditionally; there is no way of distinguishing in it analytic from other sentences because all have the same status epistemologically, all are revisable on the basis of observation. However, one must not conclude that, having rejected his earlier views on language, analyticity, etc., in 1948 Ajdukiewicz embraced a position identical with Quine's (1936: "Truth by convention"; 1953: "Two dogmas of empiricism").[24] This would not be correct. The point of introducing the concept of a language without axiomatic meaning-rules was rather to argue that the epistemological thesis of radical empiricism (according to which all science—including mathematics and logic—consists of empirically revisable sentences) and the thesis of moderate empiricism (which claims that, apart from empirical components of science, there are also *a priori* analytic ones), need not be mutually incompatible. They are not, if either they refer to different types of language or express programmes for constructing science in a certain way, rather than describe existing science. One cannot escape the feeling that this new view concerning the relationship between the two epistemological doctrines originated from the analysis of the same examples (given by Poincaré) of sentences changing status (from empirical generalisations to conventional principles). On the previous occasion Ajdukiewicz saw this change as indicating meaning and language variance, and therefore demanded that status be relativized to a language. *Now* he concluded that epistemologists only appeared to contradict one another when they attributed a different status to "the same sentence". In fact, they were talking at cross purposes, since either they had in mind different existing languages (e.g. different historical stages in the development of a scientific theory) or else were proposing or demanding that different types of language should be used in articulating science. When radical conventionalism seemed to have been undermined at the object-language level (since it was questioned whether there existed different, not-intertranslatable object-languages), it was in a sense being resurrected at the meta-level of different conceptions of language and different epistemological programmes (choice of types of language). However, if one transforms an Ajdukiewicz language based on all three types of meaning-rules into one with only deductive and empirical meaning-rules, two such object-languages are mutually non-translatable at least in their discursive parts.

This was not a transient feature of Ajdukiewicz's philosophy peculiar to his 1947 "Logic and experience". In the posthumously published paper "The problem of empiricism and the concept of meaning" (which he had read in 1962 at a conference organised in Warsaw to celebrate his 70th birthday), Ajdukiewicz distinguished the epistemological from the methodological version of the problem of empiricism. Whereas the former is concerned with *questio facti*, the latter is concerned with the question whether it is *possible* to construct science without *a priori* elements and, if it is, how should one do it? Accordingly, methodology was on this programme to study various *possible ways of doing science*, rather like geometry which studies possible spaces.[25]

This was the *coda* of Ajdukiewicz's lifelong work, the first movement of which, radical conventionalism, had been conceived back in the early thirties. Methodological pluralism, which does not preclude one from having a preferred epistemological, methodological or ideological position, has always been part of the conventionalist tradition. It seems a particularly suitable conclusion to the work of a philosopher who all his life was more tolerant towards and more ready to learn from different traditions than most of this contemporaries.

Notes

* First published in *Kazimierz Ajdukiewicz: The Scientific World-Perspective and Other Essays*, 1931–1963, XIX-LII (Edited by J. Giedymin), D. Reidel, Dordrecht, 1978. K. Ajdukiewicz was born in 1890 and died in 1963.
1. Cf. Ajdukiewicz's preface to the first volume of *Jezyk i Poznanie* [Language and Knowledge], 1960, as well as his article "W sprawie artykulu prof. A. Schaffa o moich pogladach filozoficznych" [A Reply to Prof. A. Schaff's Article Concerning my Philosophical Views], *Jezyk i Poznanie*, Vol. II, 1965, p. 176.
2. As we know, both Dedekind and Peano believed that the axioms of arithmetic *do* implicitly define arithmetical primitives. Dedekind seems to have identified the set of natural numbers with what is common to all simply infinite sets (cf. Wang, 1957: p. 152). Poincaré's argument against regarding the axioms of the Dedekind–Peano system as an implicit definition of arithmetical primitives was as follows. For an axiom system to serve as an implicit definition of its primitives it must be proved to be consistent (and categorical); non-Euclidean geometries have been proved consistent relative to Euclid's geometry and the latter relative to arithmetic; however, a proof of the consistency of the axioms of arithmetic could be only given by recursion, i.e. it would make use of the axiom of induction and would thus involve a *petitio principii*. This may be avoided only by regarding the axiom of induction as a *synthetic a priori* truth (Poincaré, 1908, 1956: pp. 151–60). It seems therefore that according to Poincaré one can prove the consistency of arithmetic only by an appeal to our mathematical intuition (if we believe in such proofs). However, if the proofs of the consistency of geometries ultimately depend on an intuitive proof of the consistency of arithmetic, one wonders why they should have a different epistemological status, i.e. why they may be regarded as terminological conventions whereas the axioms of arithmetic cannot. In any case, it is questionable whether Poincaré was right in claiming that any consistency proof of arithmetic is bound to involve a *petitio principii*. In view of Gödel's result on consistency proofs, which appeared long after Poincaré's death in 1912, he was right if by *petitio principii* is meant the use in a proof of premises "not previously justified". He need not have been right if by *petitio principii* is meant a circular proof, i.e. one in which the conclusion occurs explicitly or implicitly among the premises used. If in a proof of the consistency of arithmetic we were to use the axiom of induction the proof would not thereby be circular: the conclusion is a meta-theorem which is neither identical with nor does it imply the axiom of induction. For an analysis of the *petitio principii* along these lines, cf. Ajdukiewicz "Logic and experience" (1978: p. 169); also Ajdukiewicz's argument against the charge of the circularity of all consistency proofs of logic "Z metodologii nauk dedukcyjnych", (1921, 1966); "Axiomatic systems from the methodological point of view" (1960, 1978).
3. In his 1872 *Erlanger Programm* F. Klein defined geometry as the study of the properties of figures which remain invariant under a particular group of transformations. (This influenced H. Minkowski's space-time geometry.)
4. For each of the metric geometries there is an infinite set of different congruence classes (cf. Bonola, 1955: pp. 153–4; A. Grünbaum, 1963: Chap. I; 1968: Chaps. I and II).
5. Two non-empirical theories may be regarded as observationally equivalent. The idea of experimentally indistinguishable laws played a more important role in

Pierre Duhem's philosophy of physics (cf. his 1954: Part II, Chap. V, "Physical law").
6. The present formulation of this doctrine is mine but it is very close to Poincaré's own. It seems to imply both the "inscrutability of reference" (of the theoretical terms of a theory) and the "indeterminacy of translation" of the theoretical language of a theory.
7. Popper (1963: p. 97). For a different account of "instrumentalism" see (Giedymin, 1976: pp. 179–209, and present volume). Pierre Duhem suggested that Poincaré's account of the epistemological nature of physical theories was inspired by his study of Maxwell's theory. In "The value of physical theory" (1908, 1974: p. 319), Duhem wrote:
"... Is there a physics which has less claim to *knowledge* and which is more clearly and purely utilitarian than that English physics in which theories merely play the role of *models* without any connection with reality? Is it not that physics which first of all enticed Henri Poincaré when he was studying Maxwell's work and so inspired the famous pages in which physical theories were considered solely as convenient instruments for experimental research?..."
8. Poincaré (1905, 1958: Chap. VIII, esp pp. 98–100, 103).
9. For an earlier criticism of the analytic-synthetic distinction, see Russell (1897: pp. 57, 59): "... It will be my aim to prove... that the distinction of synthetic and analytic judgements is untenable...."; "... The doctrine of synthetic and analytic judgements... has been... completely rejected by most modern logicians...."; the logicians in question were Bradley and Bosanquet. And "... Thus no judgement *per se*, is either analytic or synthetic, for the severance of a judgement from its context robs it of its vitality, and makes it not truly a judgement at all. But in its proper context it is neither purely synthetic nor purely analytic; for while it is the further determination of a given whole, and thus in so far analytic, it also involves the emergence of *new* relations within the whole, and is so far synthetic".
10. Edouard Louis J. LeRoy (1870–1954), a student of both Poincaré and Bergson, succeeded Henri Bergson, in Collège de France and in l'Acadèmie Française.
11. Milhaud (1896); Wilbois (1899); E. LeRoy, "La science positive et les philosophies de la liberté", *Bibliothèque du Congrès Internationale de Philosophie*, Vol. I.
12. In (1906, Part II, Chap. IV) in a footnote on p. 144, Duhem refers to his article "Quelques reflexions au sujet de la physique expérimentale", *Revue de Questions Scientifiques*, 2nd series, III (1894). Cf. also *ibid.*, p. 150, footnote 7.
13. No mention has been made here of Duhem's doctrine of "natural classification" which has a more realist flavour.
14. A short account of Poincaré's philosophy of science may be found in Peter Alexander's note on Poincaré in "The philosophy of science, 1850–1910", in *A Critical History of Western Philosophy* (Edited by D. J. O'Connor).
15. These must, of course, not be confused with so-called "Ajdukiewicz categorial grammars" which originated from his "Syntactic connexion" (see Ajdukiewicz, 1978: p. 118; also Geach, 1970; and Lewis, 1970: p. 20).
16. Every sentence dictated by an axiomatic meaning-rule of an Ajdukiewicz language is an axiom in the formal sense, but *not* vice versa. An axiom of a neutral (non-assertive) axiom-system, though it is of course an axiom in the formal sense, is not dictated by an axiomatic meaning-rule in Ajdukiewicz's sense, since such a rule would require its unconditional acceptance (assertion) by any speaker of the given language. However, a neutral axiomatic system may be an Ajdukiewicz

language based on deductive and empirical or only on deductive meaning-rules.
17. Subject to qualifications (Ajdukiewicz, 1978: Section V), this point might perhaps be of some interest to those philosophers of science who—like, for example, R. Tuomela—use logical methods in discussing the semantics of empirical theories and insist on direct interpretation of theoretical terms. Cf. Tuomela (1973: p. 118).
18. This sense of "instrumentalism" was distinguished in Giedymin (1976: pp. 179–209).
19. Cf. Ajdukiewicz references to E. Spranger and his *Lebensformen*, e.g. (1960: pp. 262, 310); (1965: pp. 335). See also his reference to G. Simmel in (1978: p. 89).
20. Cf. also E. W. Beth's discussion of "A semantical version of the problem of transcendental idealism" in Chap. 22, "Realistic tendencies in the philosophy of mathematics" of (1959: pp. 619–621).
21. Ajdukiewicz also considered the pragmatic view of Ajdukiewicz languages according to which "meaning" would no longer be "ideal" and intersubjective (1978: p. 319). All these concepts, unlike the earliest one, had no epistemological consequences.
22. In "Logical comparability and conceptual disparity between Newtonian and relativistic mechanics" (1973 and Appendix, present volume), I argued that there are languages which, though not closed in an absolute sense, are closed with respect to a set of sentences. For example, the language of relativistic mechanics may be seen as closed with respect to the theoretical sub-language of Newtonian mechanics. A similar line of thought, in a much more elaborate formal framework, was pursued by Williams (1973: pp. 357–367).
23. Since in 1935 Ajdukiewicz managed to reduce empirical meaning-rules to axiomatic ones (as regards the unconditional acceptance of sentences dictated by them), the innovation under discussion here may be seen as the outcome of the process of varying the essential properties of Ajdukiewicz languages which Ajdukiewicz had begun over a decade before.
24. A suggestion to that effect seems to have been made by Skolimowski and Quinton (1973: Ajdukiewicz, 1949, Preface XVII to English edition). I criticised that suggestion in my review of their translation in *Br. J. Phil. Sci.* **25**, 3 (1974). However, it may be that their suggestion was tantamount to the statement—made in the present text—of the Quinean nature of an Ajdukiewicz language which has no axiomatic meaning-rules.
25. In "The problem of foundation", which came only one year before "The problem of empiricism and the concept of meaning", Ajdukiewicz expressed his support for empirical methodology whose aim is to study science as it actually is and has been. In the latter of the two articles he seemed to be more sceptical with regard to the feasibility of such empirical research.

5

Poincaré and the Discovery of Special Relativity

A CONSIDERABLE part of each of the four collections of Henri Poincaré's philosophical writings, *Science and Hypothesis* (1902, 1952), *The Value of Science* (1905, 1958), *Science and Method* (1908, 1956), *Last Essays* (1913, 1920, 1963), is concerned with an analysis of the geometrical and physical principles of relativity. There is no doubt whatever that these principles play an important role in his diagnostic-descriptive account of the state of theoretical physics at the turn of the century and in the associated conventionalist philosophy of geometry and physics. One is, therefore, confronted with several questions such as these. What is the historical and logical relationship between Poincaré's geometrical and physical principles of relativity and the special theory of relativity in Albert Einstein's paper, "On the electrodynamics of moving bodies" (1905, 1970)? What is the relationship between these principles and Poincaré's conventionalist view of geometry and physics? Why has this relationship never been mentioned, let alone discussed, by those philosophers of science who refer to Poincaré's conventionalism in metric geometry and physics?

A controversial answer to the first of the above questions was given by Sir Edmund Whittaker in *A History of The Theories of Aether and Electricity*, Vol. 2 (1953, 1973). In a chapter entitled "The relativity theory of Poincaré and Lorentz", Whittaker attributes the discovery of special relativity theory to Poincaré and Lorentz and gives credit to Albert Einstein only for "some amplifications" of the theory (the new formulae for aberration and the Doppler effect). Moreover, he implies that Einstein's paper "On the electrodynamics of moving bodies" (1905) was not independent of Lorentz's (1904)

"Electromagnetic phenomena in a system moving with any velocity less than that of light". Whittaker's claims flatly contradicted the established view of the matter which was held by the majority of physicists, historians of physics and philosophers of science for over thirty years, roughly between 1920 and 1953. The established view is expressed, for example, in the following passage from Hans Reichenbach's *Philosophic Foundations of Quantum Mechanics* (1944, 1965: V):

> Two great theoretical constructions have shaped the face of modern physics: the theory of relativity and the theory of quanta. The first has been, on the whole, the discovery of one man, since the work of Albert Einstein has remained unparalleled by the contributions of others who, like Hendrik Anton Lorentz, came very close to the foundations of special relativity, or like Hermann Minkowski, determined the geometrical form of the theory. It is different with the theory of quanta. This theory has been developed by the collaboration of a number of men each of whom has contributed an essential part, and each of whom in his work, has made use of the results of others....

Whittaker's controversial account of the discovery of relativity theory has had so far at least two effects, one beneficial, the other not. By challenging the established view it gave rise to a debate over the origins of relativity theory (Born, 1956; Kahan, 1959; Holton, 1960, 1964; Scribner, 1964; Keswani, 1965; Popper, 1966; Goldberg, 1967; McCormmach, 1967, 1970a, b; Heerden, 1968; Schaffner, 1969; Zahar, 1973; Miller, 1975) which has produced interesting arguments on both sides and has revealed new problems of historical or epistemological nature. Though the solution of the main issue as formulated by Whittaker remains controversial, attention has been drawn to uncontestable contributions made by Poincaré to the emergence of relativity theory. These contributions are now referred to in some textbooks of physics (e.g. Feynman, 1963: 15–3, 15–5) and excerpts from Poincaré's writings are now reproduced in anthologies concerned with the history of relativity (e.g. Pearce Williams, 1968; Kilmister, 1970). Einstein's biographers no longer ignore the influence exerted on young Einstein by Poincaré's writings (for example, Clark, 1973; and Hoffmann, 1975; in contrast to Frank, 1948). The same applies to some books on twentieth century philosophy of science. On the other hand by the apodictic style of his assertions which seemed to leave no room for doubt or argument, by an unfair assessment of Einstein's contributions (even on the assumption of Poincaré's and Lorentz's priority) and by his failure to consider the possibility of simultaneous discovery, Whittaker antagonised many and provoked some of his critics to reply in kind.[1]

I shall take as my starting point in this essay an article by Max Born, "Physics and relativity" (1956), part of which is concerned with Whittaker's views on the discovery of special relativity. I shall also refer to some of the subsequent contributions to the debate. Born's article not only initiated the debate but is still regarded as important (see, for example, Holton, 1960, 1974; Schaffner, 1969; Miller, 1975; Martin, 1976: p. 318). Its main emphasis is on the way Poincaré, Lorentz and Einstein saw—or Born thought they saw—the discovery of the "new mechanics" or of special relativity, and their respective contributions to it. The priority issue raised by Whittaker, or any other priority issue for that matter, cannot of course be resolved by considerations of this type exclusively. However, they have to be taken into account. More importantly, Born's critical analysis of the statements made by Poincaré, Lorentz and Einstein concerning the discovery directs our attention to the main problem involved in the debate, viz. *the problem of* the equivalence or otherwise of "the new mechanics" of Poincaré and Lorentz on the one hand and Einstein's special relativity on the other. It is clear that Whittaker's priority claim in favour of Poincaré and Lorentz is tenable only if Poincaré's principle of relativity is equivalent to Einstein's special principle and—more generally—if the "new mechanics" of Lorentz as improved and interpreted by Poincaré is equivalent to Einstein's special theory of relativity. In the case of non-formalised empirical theories, like the two theories in question, problems concerning intertheoretic relations are beset by difficulties such as the ambiguity or the vagueness of the concept of equivalence involved and the necessity to use roundabout methods and circumstantial evidence to establish these relations. This is especially so as the concept of the cognitive content of a theory on which the criteria of the identity of a physical theory and of the equivalence of two such theories depend, is an epistemological concept and, therefore, varies with the epistemological views of scientists or philosophers concerned.[2] This would indicate that questions of priority, in some cases at least, might be indeterminate until some epistemological criteria of content (of a theory) and hence of equivalence (of two theories) are chosen. Epistemological arguments have been introduced into the debate over the origins of relativity theory in any case by several authors, for example by Kahn (1959), Holton (1960, 1974), Goldberg (1967), Schaffner (1969), Miller (1975). The role of these arguments has been either to help in reconstructing the intended sense of various formulations of the principle of relativity in Poincaré's writings, of Lorentz transformations and of other components of the theory, or else in general to establish, hypothetically of course, the influence of Poincaré's, Lorentz's or Einstein's philosophical views on their physics.

Born's Reminiscences and Arguments against Whittaker

Max Born's challenge to Whittaker's claim is contained in Born's 1955 lecture "Relativity and physics" published in (1956) and in some of the Born–Einstein letters (Born, 1971). Judging from subsequent references, this challenge seems to carry a lot of weight. Apart from Born's authority as one of the leading physicists of the first half of the twentieth century, the following reasons must have been responsible.

At the beginning of the century, especially in 1909 and 1910, Born, then a young physicist, took part in several meetings at which "the new mechanics" was discussed by Poincaré, Lorentz, Minkowski and others. His comments on Whittaker's contention are based on his personal reminiscences of those meetings (refreshed by re-reading some of the texts) and of the general atmosphere prevailing in the physics departments in German universities at that time. In the years following 1916 Born formed a friendship with Albert Einstein whom he used to meet while they were both in Germany and with whom later on he had frequent contacts by correspondence (*The Born–Einstein Letters*, 1971, cover the period from 1916 to 1955, the year of Einstein's death). Letters which they exchanged reveal Born's deep admiration and affection for Einstein in spite of disagreements between them on fundamental questions of quantum physics. During his employment in Edinburgh University, Born—according to his own testimony (Born, 1971: p. 197)—was on friendly terms with Edmund Whittaker and argued with him about the controversial views concerning the discovery of relativity theory for three years prior to the appearance of Whittaker (1953). In other words, Born was in a perhaps unique position of having been acquainted with the four personalities crucial to the priority debate, viz. Poincaré and Lorentz on the one hand, Einstein and Whittaker on the other, and of having discussed Whittaker's contention with the latter two. Finally, the style of Born's 1955 lecture (which naturally differs from the style of his letters exchanged with Einstein) is objective and his judgement appears to be balanced.

Having outlined what seems to be relevant to the question of Born's reliability and general ability to know facts of importance to the priority dispute, one has to examine his arguments contained in the 1955 (1956) lecture. They may be divided into the following three parts.

1. A public lecture given by Henri Poincaré in 1909 in Göttingen University on the subject of "the new mechanics" (based on Book III of 1908) and subsequently published in German under the same title—a lecture which Born himself attended and the text of which he re-read for his 1955 lecture—contained a popular account of the theory of relativity without any formulae and with very few quotations. Einstein and Minkowski were not mentioned at all, only Michelson, Abraham and Lorentz:

> But the reasoning used by Poincaré was just that, which Einstein introduced in his first paper of 1905.... Does this mean that Poincaré knew all this before Einstein? It is possible, but the strange thing is that his Lecture definitely gives you the impression that he is recording Lorentz's work... (1956: p. 192).

2. On the other hand, Lorentz himself never claimed to be the author of the principle of relativity. Moreover, one of the lectures given by Lorentz in 1910, also at Göttingen University in the same series of lectures, begins with this sentence: "To discuss Einstein's Principle of Relativity here in Göttingen where Minkowski has taught seems to me a particularly welcome task". Born concludes—

"This suffices to show that Lorentz himself regarded Einstein as the discoverer of the principle of relativity" (1956: p. 192)—and points out that from the rest of Lorentz's lecture his reluctance to abandon the ideas of absolute space and time was apparent; nor did Lorentz change his position later in his life.

3. Although Einstein's (1905) paper on relativity does not contain a single reference to other people's works and thus "Gives you the impression of quite a new venture", this is not altogether true as we know from Einstein's own communication to Carl Seelig (*Technische Rundschau*, **20**, 1955) in reply to the question "which scientific literature had contributed most to his ideas on relativity during his period in Bern". In that communication Einstein stated that: (a) if looked at retrospectively, the special theory was ripe for discovery in 1905: Lorentz had observed that the transformation named after him was essential for the analysis of Maxwell's equations and Poincaré penetrated even deeper into these connections; (b) Lorentz's (1892) and (1895) were familiar at the time to Einstein but neither Lorentz's subsequent works nor Poincaré's study (*Untersuchung*) linked with the latter ("In this sense was my work of 1905 independent", wrote Einstein); (c) the new features of the (1905) paper were: the realisation that the significance of Lorentz-transformation transcends its relation with Maxwell's equations and concerns the essence of space and time as well as that "Lorentz invariance" is a general condition for any physical theory (Born, 1956: p. 194).

Critical Comments on Born's Argumentation

It would not be right to aim at Born's argument with heavy guns. In the first place, he did not pretend that it was the result of detailed and careful historical analysis of all relevant texts. In fact, having accurately recorded Whittaker's claims and references to Poincaré's writings, he himself relies on the 1909 (1908) lecture exclusively. Moreover, it is clear that his argument is not a formal, explicit one but rather of a suggestive character: there is no explicit conclusion stated and none unambiguously implied; several hints are given and the drawing of the conclusions is left to the reader. Many readers will probably conclude from what Born has said that although, in his view, it was possible that Poincaré had the same relativistic ideas as Einstein before 1905, it is also possible that, having become familiar with Einstein's relativity theory after 1905, Poincaré in 1908–9 reinterpreted Lorentz's writings in the light of that theory. However, this would be the reader's own conclusion, not Born's.

If one does draw this conclusion, then the natural follow-up along Born's line of argumentation is this: Poincaré's earlier (pre-Einsteinian) views on the subject were the same as Lorentz's (since Poincaré himself attributes them to Lorentz), but Lorentz's theory was not the same as (equivalent with) Einstein's as otherwise Lorentz would have regarded himself as the discoverer of relativity which he never did. In this way one may see the problem of the equivalence (or otherwise) between the theories of Lorentz, Poincaré and Einstein as implicit in or suggested by Born's argumentation, although the only evidence explicitly used in it consists of statements made at different times by the three parties concerned.

As a step towards an examination of the main problem of equivalence or non-equivalence of the relevant theories, let us first consider two assertions made by Born: one, that Lorentz believed Einstein to be the sole discoverer of the principle of relativity; and the other that Poincaré's reasoning in the 1909 (1908) Göttingen lecture was the same as Einstein's in (1905).

Born's inference from the fact that Lorentz used the term "Einstein's principle of relativity" in the opening statement of his 1910 lecture to the conclusion that Lorentz regarded Einstein as the sole discoverer of the principle of relativity, is not as obvious as it appears to be. Other writings of Lorentz do not all unambiguously support it. Although in both the first and second editions (1909, 1915) of *The Theory of Electrons* the credit for the *theory* of relativity is exclusively given to Einstein (the term "Einstein's theory of relativity" is used, for example in Note 72* to the 1915 edition as in the title of Lorentz's 1920 book *The Einstein Theory of Relativity: A Concise Statement* which, of course, concerns not only the special but also the general theory of relativity) and Poincaré's name appears there only in connection with the problem of the stability of the electron, this is not so in Lorentz's *hommage* "Deux mémoires de Henri Poincaré sur la physique mathématique" (1915/1921, 1954)[3]. The two memoirs of Poincaré discussed in it are the 1906 *Rendiconti* paper on the dynamics of the electron and the second of Poincaré's 1912 papers on quantum theory. Here we are interested only in the former. In the opening paragraph Lorentz emphasised that although Poincaré's memoir appeared in 1906, it had been written (and submitted, one should add) in 1905.

In order to give an appreciation of Poincaré's memoir Lorentz begins with an outline of some of the "ideas whose development resulted in the principle of relativity". Fresnel's explanation of the aberration in terms of a stationary ether is mentioned first as well as the theorem of partial entrainment of light waves by bodies in motion (Fresnel's coefficient). With the arrival of Maxwell's electromagnetic theory it was necessary to deduce Fresnel's coefficient from the electromagnetic theory. This was the problem of Lorentz's early research solved within the theory of electrons, which also explained the absence of any effect of the motion of the Earth on optical experiments, provided that one ignored terms of second order in v/c. The introduction of the (Fitzgerald) contraction hypothesis was necessitated by the negative result of Michelson's 1881 experiment which involved effects of second order. This hypothesis met with Poincaré's approval with the admonition that one should not multiply *ad hoc* hypotheses to explain particular results. Poincaré's critique encouraged Lorentz to search for a general theory whose principles would explain Michelson's and other negative results involving second order quantities. The method to be used was obvious. One had to show that the physical phenomena which occur in a material system could be described in terms of equations of the same form whether the system is at rest or in uniform translational motion. In other words, one had to find substitution (transformation) formulae for space coordinates x, y, z, and time t, and for various physical quantities such as velocity, force, etc. and to prove the form invariance of the equations with respect to those substitutions. The transformation equations for some of the physical quantities (velocity and charge) given in Lorentz (1904) were, however, not quite appropriate to that purpose with the result that the exact invariance of the equations of electrodynamics was not established.[4] The reason for this was the fact that Lorentz did not treat the reference systems x, y, z, t and x', y', z', t' on a par. He believed that in the former one operates with coordinate axes fixed in the ether and with the measurements of *true* time; the latter, on the other hand, was a "mathematical instrument" with "auxiliary quantities", among which t' or "local time" did not have the same significance as "true time". Although the difference between his transformations and those necessary to ensure the invariance of the equations of electrodynamics involved only terms which take on very small values, nevertheless—Lorentz admits—"... I have not demonstrated the principle of relativity as rigorously and universally true". The correct transformations were given by Poincaré and then by Einstein and Minkowski[5] (Lorentz, 1915/1921, 1954: p. 295; 1954: p. 685) "Poincaré... obtained a perfect invariance of the equations of electrodynamics and formulated the 'postulate of relativity', terms which he was the first to employ. In effect, by taking the point of view which I lacked, he found formulae (4) and (7) [velocity and charge transformations—J.G.]" (Lorentz, 1915/1921: p. 298). The success of relativistic substitutions was, moreover, explained in Poincaré's paper by showing that the equations of the electromagnetic theory can be cast in the form of the principle of least action and the equation which expresses that principle has the same form in systems x, y, z, t and x', y', z', t'. In conformity with the title of his memoir, Poincaré then considered the way in which the deformation of a moving electron is produced; he eliminated the hypothesis of Bucherer and Langevin (which assumes the volume of the electron to be constant) by demonstrating that it was incompatible with the principle of relativity: the principle requires that the transformations given by the formulae:

$$x' = kl(x + \epsilon t), \quad y' = ly, \quad z' = lz, \quad t' = kl(t + \epsilon x)$$

(where ϵ, k, l are constants and the units of length and time are chosen in such a way that the velocity of light $c = 1$) should form a group and this implies that l has a constant value 1. This condition is satisfied by Lorentz's hypothesis but not by the hypothesis of Bucherer and Langevin. The group of relativity consists, as Poincaré showed, of linear substitutions which leave invariant the quadratic form:

$$x^2 + y^2 + z^2 - t^2$$

Poincaré's memoir (1906) ends with an application of the principle of relativity to celestial mechanics and gravitation theory. A list of the invariants of the relativity group is given (since these alone would occur in the equations), and then it is shown that several hypotheses are admissible even if certain plausible restrictions (such as that the velocity of propagation equals that of light) are imposed. That part contains the idea of taking x, y, z and $t\sqrt{-1}$ as coordinates of a point in a four-dimensional space in which case the relativistic transformations reduce to rotations in the four-space. Similarly if one takes X, Y, Z and $T\sqrt{-1}$ as four components of force, these transform in the same fashion as x, y, z and $t\sqrt{-1}$. "I draw attention to these ideas of Poincaré"—Lorentz writes—"because they are close to those methods which were used later on by Minkowski and others to facilitate mathematical operations within the theory of relativity" (Lorentz, 1915/1912: p. 301; 1954: p. 689).[6]

If we return now to Born's argument concerning the role of Lorentz, Poincaré and Einstein in the discovery of special relativity and, in particular, to Born's account of Lorentz's perception of his own contributions, the following seems to emerge from a perusal of Lorentz (1915/1921).

It is misleading to say, as Born did, that Lorentz never claimed to have discovered the principle of relativity. Although Lorentz admitted having failed in (1904) to "demonstrate the principle of relativity as rigorously and universally true" in view of the non-symmetric (non-reciprocal) nature of his 1904 transformations, nevertheless he believed that he had discovered an "imperfect" version of that principle: his transformations established the form-invariance of the equations for charge-free space, moreover certain terms which should have vanished (but did not) from the transformation equations represented very small quantities.[7] In an effort to indicate how far his contributions in the area were overtaken by Poincaré's 1905/6 research, Lorentz writes that Poincaré "corrected the imperfections" of his work "without ever reproaching" him, made the transformations symmetric (reciprocal) by imposing group requirements on them, obtained perfect form-invariance of the laws of electrodynamics and formulated "the postulate of relativity". In other words, in his *hommage* (1915/1921) Lorentz perceived Poincaré's 1905/6 research as the attainment, with certain modifications, of one of the aims of his own research programme. The aim was to express in mathematical form (viz. that of transformation equations relative to which the laws of electrodynamics would be form-invariant) experimental relativity, i.e. persistent failure to detect by experimental means the velocity of bodies with respect to the ether. This was achieved in an equally satisfactory fashion, simultaneously and independently by Poincaré and Einstein. One cannot escape the impression that in (1915/1921) Lorentz wanted to put the record straight, as previously, in (1909, 1915), he had not given any credit to Poincaré, had emphasised "the remarkable reciprocity in Einstein's treatment of reference-frames" and had associated "the principle of relativity" only with Einstein's name.[8]

If we now assume, as is quite plausible, that Poincaré's perception of the results of his 1905/6 memoir was, on the whole, in line with Lorentz's (1915/1921), then we obtain immediately an obvious explanation of the fact which appeared strange to Born, viz. the fact that in the 1909 Göttingen lecture on the "new mechanics" Poincaré seemed to be talking about Lorentz's theory, although the central "reasoning" in it was the same as Einstein's (1905). The explanation is as follows.

The subject of Poincaré's Göttingen lecture on the "new mechanics" *was Lorentz's electrodynamics in its evolution and historical setting.* The structure of the lecture is roughly the same as that of Poincaré's 1904 St Louis lecture (1904, 1905, 1958: Chaps. VII-IX), of a part of *Science and Hypothesis* (1902, 1952: Chaps. VI, VII originally published in 1901 and with a reference to "The measure of time", 1889, Chaps. IX and X, originally 1900, Chaps. XII and XIII, originally 1889, 1890, 1902) and ultimately derives from his memoir "A propos d'une théorie de M. Larmor" (1895). From this point of view Poincaré's 1909 (1908) lecture is quite different from Einstein's (1905) relativity paper. As regards the *central reasoning* of the lecture, it is neither the same as Einstein's in (1905)—contrary to Born's claim—nor is it Lorentzian: it gives both interpretations, in the form of a disjunction, the Lorentzian as Lorentz's and the alternative ("the same reasoning as Einstein's", as Born put it) in the form of Poincaré's own comments. Apart from recording the main ideas of Lorentz's theory and their evolution in time, Poincaré introduces his own, long-held views on the principle of relativity and also indicates from that perspective the possibility of interpreting some of the concepts of Lorentz's theory in an alternative fashion. So, for example, we find statements to the effect that "the principle of relativity is a general law of nature" and has a value comparable "to that of the principle of equivalence" (1908, 1956: p. 221), that all that can be said in favour of the Lorentz–Fitzgerald contraction hypothesis is that "it is merely the immediate interpretation of Michelson's experimental result, if we *define* distances by the time taken by light to traverse them" (1908, 1956: p. 221), that "the deformation of the electrons seems extremely hypothetical" but the hypothesis may be avoided if we "imagine electrons as material points and enquire how their mass ought to vary as a function of the velocity so as not to violate the principle of relativity"; we would then see that the mass variations "must occur *as if* the electron underwent Lorentz's deformation" (1908, 1956: pp. 227–8), etc. Now the first of these pronouncements may be found in Poincaré (1895), (1902), (1905) already, whereas the succeeding comments derive from Poincaré (1905a) and (1906) as well as from his general epistemological view of the role and status of physical theories, the view according to which the descriptive content of a theory is given in terms of its observable predictions and in terms of the mathematical form of its equations. On this view the content common to two or more apparently rival theories is statable in the form of general principles. In other words, apart from recording Lorentz's electron theory and the associated theoretical interpretation of Lorentz-transformation, Poincaré pointed out the availability of an alternative interpretation of transformation equations, neutral with respect to any theoretical hypothesis. Nevertheless, since not only the electron theory on which the Lorentzian electrodynamics was based but also the anticipation, however imperfect, of the principle of relativity, were clearly due to Lorentz, and the neutral hypothesis-free interpretation of Lorentz-transformations followed from the idea of the "physics of the principles", Poincaré did not claim any credit for himself.

This account of the reasoning in Poincaré's 1909 Göttingen lecture not only gives an adequate answer to the question posed by Born but also is the only one compatible with available evidence, in particular with the statements made by Lorentz, Poincaré and Einstein concerning their respective contributions to relativity theory. Moreover, it can be easily reconciled with Whittaker's position as soon as one distinguishes between the non-mathematical formulations of the principle of relativity (it is impossible to detect absolute motion, or motion relative to the ether, by any physical experiments whatever; all physical laws remain unchanged whether referred to a stationary frame or to one in uniform, rectilinear motion), for which priority should be given to Poincaré, and the mathematical formulation (with applications) given simultaneously and independently by Einstein and Poincaré.

There is, however, another possible approach to the issue. Any similarity between Poincaré's and Einstein's reasoning would be declared, from this point of view, as purely formal (Poincaré's 1906 memoir is then dismissed with the compliment that it is "an impressive piece of mathematics"); Poincaré's indication of the availability of the hypothesis-free interpretation of Lorentz-transformations (and of the principle of relativity) would be either ignored or pronounced as too weak, "half-hearted", "short of commitment", etc. The continuity between Poincaré's research and Lorentz's "electromagnetic world-view" would be alleged and Poincaré's theory would be claimed to be an improved version of Lorentz's theory, hence not equivalent to Einstein's. We shall consider this line of argument because it has proved very popular in the debate over the origins of relativity theory, and because its examination will show whether the view outlined in the present chapter can be maintained.

Observational versus Conceptual Equivalence

The line of argument, then, to be considered now is this.

One aims at establishing the non-equivalence, in some sense, of Lorentz's (1904) electrodynamics and the electrodynamics based on

Einstein's (1905) special relativity. The next step is to argue that Poincaré's 1905/6 research on the dynamics of the electron, together with its postulate of relativity is a mathematically improved version of Lorentz's (1904) theory, hence it is "continuous" (in this sense equivalent) with the latter.

If successful, this argument would not only refute Whittaker's thesis of Poincaré's priority, but also the thesis, mentioned in the previous chapter, of the simultaneous discovery of (special) relativity by Poincaré and Einstein coupled with the claim of Poincaré's priority in formulating the principle of relativity in the non-mathematical form. However, for the argument to be successful in this sense two conditions would have first to be satisfied. The nature of the relevant equivalence relation(s) would have to be made sufficiently clear so that there is consensus as to what constitutes evidence for or against each of the premises. Moreover, whatever other properties it might have, that relation would have to be a proper equivalence, i.e. to be a transitive relation. It is not at all obvious that the second of these conditions, necessary for the validity of the argument (the conclusion being, of course, the denial of the equivalence between Poincaré and Einstein), is or can be satisfied. On the contrary, the argument appears to rely on more than one concept of equivalence and one of these (theory T_2 is continuous with theory T_1) is rather like "x is similar (in certain respects) to y", which is not transitive. To illustrate some of the difficulties arising from the fact that one is not dealing here with purely mathematical theories, let alone formalised ones, and does not operate with a formal concept of equivalence, let us consider the case in which an attempt is made to establish (partially at least) the non-equivalence of Lorentz's and Einstein's theories by showing that the two theories were seen as rivals "for a number of years" after 1905. This is one of Schaffner's arguments in (1969: p. 510) in which the evidence adduced is taken from the writings of M. von Laue, N. R. Campbell and P. Ehrenfest (the first two dated 1911). Here is a passage from Campbell as quoted by Schaffner:

> ... It is important to notice that there is another theory, that of Lorentz, which explains completely all the electrical laws of relatively moving systems; that the deductions from the Principle of Relativity [Einstein's—J. G.] are identical with

those from the Lorentzian theory, and that both sets of deductions agree completely with all experiments that have been performed. If then, anyone prefers one theory to the other it must be either on the ground of differences in the laws not contemplated originally which are predicted respectively by the two theories or because of some general grounds independent of experimental considerations....

It is clear from the quoted passage that Campbell regards the theories in question as observationally equivalent, at least as far as both theories have been actually developed, i.e. whatever observational consequence has been derived from Lorentz's theory is also derivable from Einstein's and vice versa. They could be rivals only with respect to possible observational consequences as yet not discovered, should such consequences be in fact obtained and prove to make a difference between the theories. In such a case the theories would turn out not to be observationally equivalent, hence simply *not* equivalent. But the possibility is not excluded that they are observationally equivalent in the strict sense, in which case they would have identical content in the sense of Ramsey (see Chap. 2, this volume), i.e. would be—in his sense—equivalent.

A similar view was expressed by Moritz Schlick, perhaps the first German-speaking philosopher of some note (and with a physicist's background) to apply his talents to the analysis and popularisation of the epistemological assumptions and consequences of Einstein's relativity theory whose efforts in that direction, moreover, seem to have had Einstein's approval.[9] Schlick makes two points, one of which concerns the observational and mathematical equivalence of Lorentz's and Einstein's theories, the other their conceptual-theoretical disparity or non-equivalence:

... The mathematical form of the laws governing the course of natural processes is exactly the same for both theories; only its *interpretation* is different. Thus both theories do the same thing, but Einstein's is very much simpler, since it employs only a single explanatory principle, whereas the other needs a series of special hypotheses....

What does Schlick mean by "interpretation" with respect to which Lorentz's and Einstein's theories are supposed to differ? It is the conceptual (and ontological) disparity, associated with the greater simplicity of Einstein's framework. Whereas according to Lorentz

"... a real influence of absolute motion *does exist*, but besides that a series of other physical influences is hypothetically assumed in order to explain why this influence is not observed", Einstein does not have to introduce *ad hoc* hypotheses to establish agreement with experience, since according to his theory no absolute motion, hence no influence of absolute motion, exists (Schlick, 1915, 1979: p. 161). The conceptual difference indicated is also symptomatic of the difference in "the line of thought leading to the equations" (Schlick, 1915, 1979: p. 164). In other words, Schlick's account of the relationship between Lorentz's and Einstein's theories allows for their observational equivalence (i.e. they explain the same facts, yield the same predictions), for the laws and transformations having the same mathematical form (presumably if one disregards the "imperfections" in Lorentz's 1904 transformations) and—at the same time—for their conceptual disparity. The last of these features is due, according to Schlick, to the difference in the line of thought leading to the discovery of the same equations: whereas Lorentz's thinking relied on the belief in the existence of the ether as the medium in which electromagnetic effects are propagated and which forms an absolutely stationary reference frame, Einstein's point of departure was the denial of the existence of absolutely stationary media or bodies, of the existence of absolute motion, velocity, hence of any effects of these. The theoretical-conceptual disparity between these two mathematically and observationally equivalent theories is due, on Schlick's account, to the fact that their ontological assumptions are mutually contradictory.[10]

The passages quoted from Campbell's and Schlick's writings amply show the ambiguity of "equivalence" used in the discussions and controversies concerned with inter-theoretic relations. They also show that the same pair of theories may be equivalent in one sense and non-equivalent in another. If one puts emphasis on the conceptual disparity between the abstract parts of Lorentz's and Einstein's theories, then one will deem them non-equivalent (conceptually). If, on the other hand, one sees these differences as of secondary importance from the natural scientist's as opposed to a historian's point of view, then one will insist that the theories are equivalent (observationally). A good example of the latter approach (which shows by the way that there were also scientists—disregarded perhaps by Schaffner (1969)—who for several years after 1905 did *not* see Lorentz's and Einstein's theories as rivals), is W. de Sitter's "On the bearing of the principle of relativity on gravitational astronomy" (1911). De Sitter opens his article by saying that the principle of relativity, although first developed in connection with the electromagnetic theory of light, has been recently more and more considered as of universal application; his paper would discuss its consequences for the laws of planetary motion. For the physical meaning of the principle of relativity de Sitter refers readers to papers by H. C. Plummer (1910) and E. T. Whittaker (1910) adding, however, the following footnote (1911: p. 389):

> ... Both authors [i.e. Plummer and Whittaker—J.G.] make free use of the word "aether". As there are many physicists nowadays who are inclined to abandon the aether altogether, it may be well to point out that the principle of relativity is essentially independent of the concept of an aether, and, indeed, is considered by some to lead to a negation of its existence. Astronomers have nothing to do with the aether, and it need not concern them whether it exists or not. All Mr Plummer's results remain true, and retain their full value, if the "aether" is eliminated from his terminology. And also in Mr Whittaker's note the word "aether" is not essential, except, of course, from an historical point of view....

It will be useful for the main problem of the present chapter to record also the following statements by de Sitter: "... The starting point of my investigations has been the papers by Poincaré and Minkowski.[11] The manner in which the equations of motion are derived below is entirely derived from the last section of Poincaré's paper. I also owe much to conversations with and advice from my colleague Professor Lorentz..." (de Sitter, 1911: p. 389). There is no mention of Einstein's name in de Sitter's paper (1911), even though there is a reference (p. 394) to "Lorentz's Göttingen Lectures, October 1910 (*Physikalische Zeitschrift*, vol. XI)" which was the starting point of Max Born's reminiscences discussed in the previous chapter. This omission is presumably due to de Sitter's belief (at the time) that Einstein's paper on relativity (1905) contained a theory equivalent with the combined content of Lorentz's (1904) and Poincaré's "December 1905", i.e. (1906) and the latter had the advantage of providing a chapter in which the principle of relativity was applied to celestial mechanics and gravitation theory, the area with which de Sitter's paper is concerned.[12]

It will be instructive to mention also one more contemporary paper on the subject, this time by a mathematician: Felix Klein in "Über die geometrischen Grundlagen der Lorentzgruppe" (1910) looks at relativity theory from the point of view of Poncelet's and Cayley's projective approach to metric geometry. "... the new developments in physics which to the layman so easily give the impression of something paradoxical, turn out to be a corollary—so to say—of a general, for quite a long time well-systematised way of thinking...." (Klein, 1910: p. 281). One could identify the concept "the theory of invariants relative to a transformation group" with the concept "relativity theory with respect to a group". "What is called by modern physicists 'relativity theory' is the theory of invariants of the four-dimensional space–time x, y, z, t (the Minkowskian 'world') relative to a certain group of collineations, viz. the Lorentz-group" (Klein, 1910: p. 287). Moreover, "... in the four-dimensional world, the system of mechanics is subsumed under the concept of the projective approach to metric geometry, and this applies to both the system of classical mechanics as well as the *new* mechanics of Lorentz, Einstein, Poincaré and Minkowski; the nature of both systems and especially their mutual relationship could in this way be best clarified" (Klein, 1910: p. 293). The difficulties which one experiences in studying non-Euclidean geometries when starting from the Euclidean point of view are similar to those one comes across in approaching the new mechanics from the point of view of the classical system. These difficulties would disappear if in the latter case one adopted instead the more general affine standpoint (it is not necessary to adopt the even more general projective one) and used the associated invariant theory, already elaborated by mathematicians, to treat the classical and the new mechanics as special cases (Klein, 1910: p. 299).[13] As we see, Klein emphasised not only the continuity of Lorentz's, Einstein's, Poincaré's and Minkowski's contributions to relativity but also the comparability of classical and relativity mechanics as special cases of the invariant theory of the affine group (i.e. the relativity theory of the affine group).

However, between (roughly) 1920 and 1953 the view became firmly established according to which relativity theory was a system of revolutionary scientific ideas which differed radically from the classical system and, moreover, was the product of one man. This view was expressed in a passage from Reichenbach's book quoted at the beginning of the present essay.

When in the early nineteen-sixties the problem of the origins of relativity, revived by Whittaker (1953), began to attract attention of both historians and philosophers of science, this coincided with and was stimulated by the new wave of anti-positivist and anti-empiricist tendencies among many outspoken philosophers of science, tendencies which favoured a historist, psychologist or sociologist approach to the philosophy of science in place of empiricist epistemology and logical analysis. Content and meaning were not to be understood in terms of theory-independent observational consequences but in terms of "research aims" and "programmes", "ontological commitments", "conceptual frameworks", "theory-impregnated perceptions", etc.[14] Significantly, one of the first and most influential contributions to the debate at that time, Holton's "On the origins of the special theory of relativity" (1960, 1974: pp. 165–84) begins with an emphatic rejection of the doctrine attributed to logical empiricists that the philosopher of science "... is not interested in the context of discovery, but in the context of justification".[15]

If attention is focused exclusively on the context of the justification of theories through predictions they yield, then one is naturally led to the identification of the meaning or content of a theory with the class of its observational consequences and to the idea of the equivalence of two theories in terms of observational equivalence. If, on the other hand, the context of discovery is taken into account, then science will be seen, like any other human activity, as arising from some tradition (whether in the sense of its continuation or of breaking away from it) and as goal-directed. Meaning is then linked to humanistic understanding.[16] If two theories in a given area of science originated from very similar traditions, and are linked with very similar cognitive commitments and aims—the last two not necessarily consciously recognised—then they are equivalent in meaning. This socio-historical meaning-determination applies not only to scientific theories when examined by historians or historically minded philosophers but also—at the (meta-)level—to the writings of historians and philosophers of science. Their socio-historical perspectives interact with the material they study and may have to be unmasked (de-mystified) if they produce distortions (not necessarily intended).[17]

The Question of Equivalence: Lorentz's Research Programme and Poincaré's Epistemology

Whether or not one approves of hermeneutics as a foundation of semantics, philosophy of language, epistemology and philosophy of science, there is no question that insofar as its methods are part and parcel of historical analysis they are legitimate in the context of the research into the origins and discovery of relativity theory. Naturally this does not mean that *any* conclusion reached in this way is acceptable.

What does it mean to claim that Poincaré's research in theoretical physics in the period roughly between 1895 and 1906 was part of, or continuous with, Lorentz's research programme? It means, among other things, to claim that Poincaré subscribed to: (1) Lorentz's electromagnetic world-view based on the electron theory of matter, with the ontological implication that the ether and electrons are the ultimate constituents of the universe, and associated with the programme of reducing all physics to the electron theory; (2) the belief that the stationary ether, as the medium of the propagation of the electromagnetic effects, provided a privileged frame of reference with respect to which bodies may be absolutely at rest or in absolute motion, the time measured is the true time and the velocity of light assumes the constant value c; (3) the explanation of the failures to measure the velocity of the Earth through the ether in terms of various compensatory effects such as the real, hence absolute (Fitzgerald–Lorentz) length contraction of all bodies (including electrons) in the direction of their motion through the ether; (4) the use of non-reciprocal Lorentz transformations which make provision for the ether as the privileged frame and which do not, therefore, form a group and do not ensure strict covariance of the laws of electrodynamics ("v" in these transformations stands for "the velocity relative to the ether" and "t'" for "local time", "t" for "true time" and the length contraction by the factor $\sqrt{1 - v^2/c^2}{:}1$ is interpreted as real, physical contraction due to the motion of bodies through the ether); (5) the restriction of the new transformations to the laws of electrodynamics and the retention of the Galileo–Newton transformations for mechanics (hence also of the classical theorem for the addition of velocities of material bodies); finally (6) the naïve realist view of a physical theory which postulates the ether and electrons as "the only true matter" behind the phenomena and which makes use of metaphysical concepts allegedly referring to quantities which in principle can never be measured (e.g. "the velocity through the ether", "length contraction of all bodies moving through the ether", etc.).

In the debate over the origins of relativity theory most critics of Whittaker's thesis have claimed that all the mentioned points (1) through (6) may be found in Poincaré's writings. For example, Holton in (1964, 1974) writes that Poincaré "... remained unshakeably against Einstein's interpretation to the end" (p. 187), that "the existence of the ether is rarely doubted" in Poincaré's writings, "... for, like Lorentz, Poincaré explained by compensation of effects the apparent validity of absolute laws in moving inertial systems, and maintained the privileged position of the ether" (p. 187), the main difference between Einstein and Poincaré being that "... Einstein had fully embraced relativity, by elevating what he called his *Vermutung* (conjecture) namely, the Principle of Relativity, to the status of a *Voraussetzung* (postulate)..." whereas for Poincaré the principle of relativity was an experimental fact (pp. 189–90) which became suspect as soon as W. Kaufmann's 1906 experiments appeared to undermine it (p. 189). In an earlier paper (1960, 1974: p. 176) Holton claimed that the principle of relativity in Poincaré's 1904 St Louis lecture (1905), which Whittaker quotes as Poincaré's anticipation of the special principle of relativity, was the Galilean–Newtonian principle and not a new principle at all. Similarly, Miller in his detailed and otherwise most instructive study (1975) of Poincaré's (1906) memoir attributes to Poincaré the metaphysical view that "... Beyond the electrons and the ether there is nothing...", quoted from

Poincaré's "The new mechanics" (1908, 1956), (the same is repeated by Goldberg, 1967: pp. 938–9). Miller concedes that Poincaré seemed to have all the requisite concepts for a relativity theory, nevertheless Poincaré's "relativity theory was to be an inductive one with the laws of electromagnetism as the basis for all physics", this "prevented him from understanding the universal applicability of the principle of relativity and therefore the importance of the constancy of the velocity of light in all inertial frames... (Miller, 1975: pp. 319–20).

Miller sees as the main result of Poincaré (1906) the proof that of all the existing theories of the electron only Lorentz's was compatible with the principle of relativity provided one introduced forces of non-electromagnetic origin to hold together the deformable electron. Neither Abraham, Lorentz nor Poincaré, according to Miller, could perceive the larger issue of a universal theory of relativity: "This theory could be obtained only by deductive method. To the extent that Abraham, Lorentz and Poincaré were committed proponents of the electromagnetic world-picture, they could neither discover nor accept the universal theory of relativity..." (1975: p. 234).

S. Goldberg (1967) repeats Holton's assertion that Poincaré's principle of relativity had a different epistemological status from Einstein's (hence the two were not equivalent) and links it with Poincaré's "inductivist" method of doing theoretical physics (in spite of his conventionalist philosophy) and with his participation in Lorentz's electromagnetic research programme (Goldberg, 1967: pp. 938–9). Goldberg claims, moreover, that whenever Poincaré refers to the constancy of light velocity (as, for example, in his 1898 paper on the measurement of time) he means constancy "with respect to the ether", as implied by Maxwell's equations (Goldberg, 1967: p. 939).

In order to emphasise more clearly the difference between Lorentz's programme on the one hand, which Poincaré was supposed to follow, and Einstein's on the other, Schaffner (like many other participants in the debate) refers to a distinction—which he believes was due to Einstein—between constructive theories and theories of principle. Lorentz's electrodynamics was a constructive theory since it was based on the electron theory which postulates an electron structure of matter to explain the phenomena. Einstein chose to approach problems through a theory based on "a universal formal principle" (i.e. the principle of relativity) because, as he reported in his (1905a), Maxwell's theory and the electron theory failed completely for black body radiation at short wavelengths and low radiation densities and he despaired of "discovering the true laws by means of constructive efforts based on known facts" (quotation from Einstein's Autobiographical Notes, Schilpp (1949) given in Schaffner, 1969: p. 513).

Some of the cited commentaries on Poincaré's contributions to relativity theory are too obscure to be properly evaluated. This applies, for instance, to the statement that "a universal" theory of relativity, such as Einstein's special theory, could be obtained only with the help of "deductive method" whereas Poincaré, allegedly, was determined to base his on "induction" (Miller, 1975). In some of the other commentaries terminological misunderstandings appear to be responsible for rather strange conclusions as when, for example, it is argued that since for Poincaré the principle of relativity was an "experimental result", it could not possibly function as a postulate in a theory (Goldberg, 1967). It can be easily verified that Poincaré referred not only to the principle of relativity but also to other so-called principles of physics (the principle of conservation of energy, of least action, of action and reaction, Carnot's and Hamilton's principles, etc.) as experimental facts (truths, results), there seems to be no earthly reason why experimental results *in this sense* should not be postulates in respectable theories; as a matter of fact, some of them are. These odd claims about Einstein's "deductive method" and Poincaré's "inductive" inclinations have some family resemblance with Holton's arguments concerning the difference in the epistemological status between Einstein's and Poincaré's principles of relativity (and their consequent non-equivalence in a sense), arguments which are based on Holton's own idea of the "thematic" elements and origins of science. However, Holton's suggestion that by elevating the principle of relativity (as a conjecture) to the status of a postulate Einstein (1905) turned the principle into a thematic proposition, neither verifiable nor falsifiable but tenable "come what may", does not seem to be compatible with Einstein's own numerous expositions of his theory in which experimental evidence in favour of the principle is adduced.

From the point of view of the dispute over the origins of relativity theory, the most important contention in the quoted commentaries is Poincaré's participation in and adherence to Lorentz's electromagnetic research programme. Many of the claims to that effect seem to have been made without any clear evidence adduced in their support. Where, for example, is the evidence to substantiate the claim (Miller, 1975) that in Poincaré's relativity theory the laws of electromagnetism were to be the basis for all physics (presumably in the sense of the reduction of all laws of physics to the electromagnetic theory)? In other cases evidence is used uncritically, and easily available *prima facie* contrary evidence is ignored or dismissed (as in Holton's, so to speak, frequentist interpretation of Poincaré's ontological commitments: "the existence of the ether is rarely doubted [in Poincaré's writings—J. G.]"). The most striking examples are perhaps among the arguments purporting to establish that Poincaré was a firm believer in the existence of the ether as a substance. In suport of this claim one quotes, for instance, the statement "Beyond the electrons and the ether nothing exists" from Poincaré's "The new mechanics" (1908, 1956: p. 209).

Now the statement in question occurs in the third paragraph of the $2\frac{1}{2}$-pages-long Section IV entitled, quite clearly and explicitly, "Lorentz's theory". Section IV is a bare report of Lorentz's theory of electrons and there should be no doubt in any reader's mind that the statement is there to convey one of Lorentz's ontological views, not Poincaré's. Again, Poincaré's concern that the principle of relativity might be falsified by certain experimental findings (Kaufmann's 1906) is emphasised in order to contrast it with Einstein's almost dogmatic attitude towards the principle (hence its "thematic" status in Holton's 1974 p. 190) *but* a similar concern (expressed quite unambiguously in Poincaré, 1906: Introduction) that Lorentz's whole (entière) theory might be endangered by the discovery of cathode rays did not raise any doubts as to whether Poincaré was a firm supporter of Lorentz's research programme (with all *its* "thematic" elements).

Finally, one quotes (Miller, 1975: pp. 236–7) Poincaré's statements such as that "the origin of our belief in the ether" is to be sought in the need to postulate a material agency for the propagation of light

(Poincaré, 1902, 1952: p. 169), or that "... the ether is no less real than any external body" (Poincaré, 1905, 1958: pp. 139–40) in order to establish Poincaré's commitment to Lorentz's ontology. However, no mention is made of other passages (in the same works of Poincaré) which would affect our understanding of the previous ones. So, for example, Poincaré explained that "the ether has been invented" so that the laws of ordinary mechanics could have the form of differential equations rather than of equations of finite differences (Poincaré, 1902, 1952: pp. 237–8). Moreover, he stated repeatedly that the question whether ether exists is a metaphysical one and should be left to the metaphysicians (Poincaré, 1902, 1952: pp. 211–2). Even Fresnel, according to Poincaré, was not at all concerned with that question but rather with predicting optical phenomena (Poincaré, 1902, 1952: p. 160).

To conclude, though there is little doubt that critics and opponents of Whittaker's account of the discovery of relativity have in their historical studies contributed much to a better understanding of the relationship between Lorentz's, Poincaré's and Einstein's relativity theories, nevertheless their handling of the hypothesis crucial to the debate, viz. of the hypothesis that Poincaré was an uncritical adherent of Lorentz's theory and research programme, is objectionable on both historical and methodological grounds. In the first place, in their use of textual evidence they have failed systematically to distinguish between (1) what Poincaré merely *reports* as Lorentz's theory(ies), as in Sect. IV of "The new mechanics", (2) what he *assumes*, from Lorentz's theory for argument's sake, as in his contribution to the problem of the stability of the deformable electron in the (1906) memoir, and (3) what—if anything—he *affirms* as his own views or beliefs. Had they distinguished between (1), (2) and (3), several of their arguments in support of the hypothesis would not be valid. Secondly, the hypothesis is almost impossible to reconcile with the conventionalist philosophy Poincaré professed, with the "inductivist" philosophy the critics attribute to him or with the mathematical tradition in the light of which Poincaré himself saw his role *vis à vis* physics. And yet no alternative hypothesis, free from these shortcomings, has been considered. Nor has any serious attempt been made at a sympathetic understanding of Poincaré's position.

Such an alternative hypothesis will be outlined presently and it will be claimed that it is in better agreement with textual evidence, than the one in terms of Poincaré's (alleged) adherence to the Lorentzian research programme. The alternative may be seen as a reconstruction of Poincaré's situational logic, or as a reconstruction of the research programme which he supported, and consists of four points.

(a) Poincaré did not see his role *vis à vis* physics as a physicist who invents new theories or hypotheses for experimental testing; rather he saw his role, whether in the area of celestial mechanics and astronomy or in the area of theoretical physics, as a mathematician in the tradition of men like Euler,[18] Laplace, Lagrange, Hamilton, Fourier, etc., but especially those who contributed to the mathematical development of "the physics of the principles".

(b) In the history of modern physics Poincaré distinguished three stages, the first of which—the physics of central forces—was dominated by the desire to discover the ultimate constituents of the universe and the hidden mechanism behind the phenomena; the second—the physics of the principles—aimed at formulating mathematical principles which systematise large areas of empirical facts and are neutral with respect to different theoretical interpretations but susceptible to any of them; the third—at the turn of the century—whose image and future was not quite clear yet but which was characterised, among other things, by the criticism of some of the principles of the previous periods, by the revival of the search for ultimate constituents (electrons), by the importance of statistical laws, etc. (Poincaré, 1905: Chaps. VII-IX).

(c) The main epistemological problem which emerged for Poincaré from the evolution of modern mathematics and physics was the problem of objectivity and rationality of science. How is objective scientific knowledge possible in the face of frequent and apparently disruptive changes in mathematics (e.g. the discovery of non-Euclidean geometries, of many-dimensional spaces, etc.) and theoretical physics (the ephemeral character of physical theories at the turn of the century), changes which are not explainable either in terms of the nature of our mind or in terms of logic and experience? (Poincaré, 1902: Preface, Chap. X).

(d) Poincaré's solution of the problem of change and objectivity, embodied in his conventionalist epistemology, relies mainly on the philosophy of the physics of the principles and on the ideas suggested by the theory of invariants with respect to transformation groups. It implies that frequent changes, not dictated by experimental findings, affect conventional components, which give rise to different but equivalent systems of measurement and different but observationally equivalent and structurally similar theoretical descriptions. A change of conventions which preserves the observational (numerical) content and the form (type of structure) of a theory is like a change in the coordinate system or like a change from one to another equivalent reference frame. The conventionalist solution implies a limit of theoretical knowledge: two theories which make no observable difference and postulate similar mathematical structures are indistinguishable.

What kind of attitude towards theories in general and Lorentz's theory and the electromagnetic "world-picture" in particular would one expect on the basis of (a) through (d)?

One would expect Poincaré to *examine* critically, *compare* and *evaluate* existing theories in physics to *distinguish* their common *empirical content* from more speculative, theoretical components with respect to which they differ, to suggest or devise *mathematical principles* which would systematise the common empirical content, to examine whether a *dictionary* could not be found to *translate* apparently rival theories or *correspondence* established between the *structures* (models) they postulate, whether one of the theories could not be seen as a *limiting case* of another or whether two or more competing theories could not be shown to be *special cases* of a more general, abstract theory. One would, moreover, expect Poincaré to *contribute* to the most promising theory by providing solutions to some of its mathematical problems, by suggesting or devising a new formalism, or alternative interpretation of its elements, by revealing formal analogies and similarities with other areas, etc. One would expect him to affirm, if anything, only the theory's empirical laws and empirical predictions but not its speculative-theoretical assumptions. So, for example, he would assert the principle of relativity as the claim that it is impossible by any physical means to measure the absolute velocity of the earth (or as the statement to the effect that the laws of physics are form-invariant with respect to a group of transformations, e.g. the Lorentz group), as long as this is in conformity with experimental findings. On the other hand, he would neither affirm (as his belief) nor deny that beyond the ether and electrons there is nothing. With regard to the hypotheses (and the concepts) of the ether, contraction, etc., one would expect him to examine their function within the theories in question, trace their links (if any) with experimental data and with formal features of the theory for which they might be responsible and, finally, consider or suggest interpretations alternative to the original ones. If one reads Poincaré's writings on the physics of his time, it seems that this is exactly what he was doing. A brief survey of some of his writings between 1890 and 1905 may help to bring this to our attention.

Although it is usually taken for granted that Poincaré's conventionalist philosophy concerns exclusively the epistemological foundations of metric geometry (the conventionality of the metric) and, indeed, his epistemological views were expressed for the first time in print in the memoir "Sur les hypothèses fondamentales de la geométrie" (1887), nevertheless it seems that some of the essential ideas of conventionalism, such as the relativity of ontology based on what Poincaré called "parallelisms", relevant not only to metric geometry but to other areas of mathematics and physics, the associated inter-translatability of languages, etc., occurred to him in the context of his discovery of automorphic functions (1881, and about eleven memoirs between 1882 and 1898; see Chap. 2 in this volume and Ford, 1929, 1951). For he established then connections between hyperbolic geometry and certain automorphic functions (more generally, the isomorphism between the group of hyperbolic movements of space and the group of proper linear substitutions of a complex variable) and also between groups of linear fractional transformations and the solutions of linear homogeneous differential equations. Between 1892 and 1899 he introduced, on the basis of his earlier mathematical investigations, new formal methods into celestial mechanics. As a mathematical model of a dynamical system he used a field of vectors on a phase space, the latter being of non-linear type (such as a generalised cylinder) rather than Euclidean (Poincaré, 1892-99). Between 1891 and 1900 Poincaré extended his epistemological analysis to the foundations of geochronometry, kinematics, mechanics and (from 1895) to optics and electrodynamics (1902, 1952: Chaps. III-X and XII-XIII). Essential to his analysis were the concepts of the relativity or conventionality of space and geometry.

Space, considered as a mathematical continuum in abstraction from measuring instruments, has only topological properties; it has neither projective nor metric ones. As such it is amorphous in the sense that it does not differ from any space into which it could be continuously deformed (relativity of space in the wide sense; Poincaré, 1905, 1958: p. 39; 1913, 1920, 1963: p. 27). This explains why—if the nature of space and of geometry are properly understood—the discovery of non-Euclidean and of higher-dimensional geometries should not only sustain our trust in the continuity of mathematical thought but inspire our admiration for its creative spirit. Metric properties of figures are neither imposed *a priori* on our minds— contrary to Kant's philosophy—nor—contrary to geometric empiricism—are they intrinsic properties of some absolute space—to be discovered experimentally, by direct measurement (for example, of the parallaxes of distant stars) or by the "violation of the law of relativity of space" (which states that the laws of phenomena in a material system depend on the state of the bodies in the system and on their mutual distances, but do not depend on the absolute position or orientation of the system, or in other words: "The readings that we can make with our instruments at any moment will depend only on the readings that we are able to take on the same instruments at the initial moment" (Poincaré, 1902, 1952: p. 77)). The results of measurements do not give us access to space, they merely express relations between our measuring instruments (i.e. rigid rods and light rays) and other bodies. The conventionality of metric geometry can be expressed by saying that space may be metrised in more than one way consistent with its topology. In another terminology, the principle of relativity of metric geometry states: there is a group of transformations which conserve some properties of figures (which we call the shape or form of geometrical figures relative to the given group). No group is specified in this principle so that in effect it is compatible with the choice of an arbitrary group with the restriction that it must not transform a figure into its parts. Two groups with a similar structure are indistinguishable and two languages or systems of concepts associated with such groups are intertranslatable; such a dictionary for the Euclidean and Lobatchevskian geometries is provided by Darboux transformations (Poincaré, 1887; 1891, 1902: Chap. III; 1898, 1905: Chap. II; 1898a; 1902, 1952: Chaps. IV, V). The dimensionality of space depends on the choice of the space element (point, line, plane, etc.) as has been shown by Plücker (Poincaré, 1898a). When we say that our world has three dimensions, we mean this: to preserve predictive "parallelism" with Newtonian physics, which assumes three-dimensional space (and one-dimensional time) any set of laws reformulated in terms of an n-dimensional space, where $n \neq 3$, would have to be a set of laws invariant with respect to a group isomorphic with the Newtonian group (Poincaré, 1913, 1920, 1963: pp. 39–42).

Geo-chronometry is the basis of *mechanics*. Classical mechanics is formulated as if facts could be referred to *absolute space* and *absolute time*. However, experiments teach us only the relations of bodies to one another and we have no intuition of "the simultaneity of two events occurring in two different places"; the statement that two such events are equal can only acquire a meaning by a convention. In other words, "absolute space", "absolute time" and "Euclidean space" are only conventions on which classical mechanics is constructed and which could be changed (Poincaré, 1897; 1902, 1952: p. 90). The relativity of motion is acknowledged in classical physics in terms of *the principle of relative motion* according to which "The movement of any system whatever ought to obey the same laws, whether it is referred to fixed axes or to the movable axes which are implied in uniform motion in a straight line". This principle is imposed upon us for two reasons "the commonest experiment confirms it; the consideration of the contrary hypothesis is singularly repugnant to the mind" (Poincaré, 1902, 1952: p. 111). However, the principle does not appear to be true for accelerated motions or rotations, at least it does not seem to be in conformity with experiments as they are usually interpreted. It is from such experiments that Newton concluded the existence of absolute motion. If, however, we find the very idea of absolute space meaningless (geometrically and physically speaking), we can neutralise it by seeing in it *an artifice*. "Absolute space" as such a conventional artifice may be eliminated if—by differentiation we generalise the laws of mechanics and express them in the form of third order differential equations rather than second order, as at present; all frames, whether inertial or not, would then be equivalent. "... the principle of relativity will still be applicable in this case. When we proceed from fixed axes to rotating axes these equations of the third order will not vary. What will vary will be the equations of the second order which define the coordinates...." (Poincaré 1913, 1920, 1963: p. 21; 1902, 1952: pp. 78-9, 112-22).

In a review of Hertz's *Principles of Mechanics* Poincaré (1897; 1954: pp. 231-50) gives a critical comparison of the classical formulation of mechanics (based on the concepts of space, time, mass and force, and on Newton's laws), the "energetist" formulation (based on the concepts of kinetic and potential energy, on the principle of conservation of energy and the Hamiltonian principle), both criticised by Hertz in his book, and the Hertzian formulation which assumes the hypothesis that mechanical processes proceed as if the connections between the various parts of a physical system acting one upon another were fixed and on the hypothesis that there are invisible masses with concealed motions, responsible for observable motions. Since Hertz's theory may be seen as a kind of generalised kinematics with part of the system under consideration concealed and part observable, in his discussion of Hertz's hypotheses, Poincaré quotes a theorem of Gabriel Koenigs on "articulated systems". An articulated plane system is a system of slabs or simple plane figures some of which are connected among themselves by pivots normal to the plane. The movements of articulated systems are studied with the help of the theory of geometric transformations. Among theorems proved by Koenigs (originally in 1895) is the following

> Any algebraic curve or surface may be described with the help of an articulated system. Or, more generally, given any algebraic relation there is an articulated

system such that in virtue of its connections the coordinates of the various points of the system satisfy that algebraic relation (Koenigs, 1897: pp. 302-5; Poincaré, 1897; Tome, 1954: IX pp. 248-9).

The mentioned theorem and Koenigs's kinematics in general played an important role in the formulation of Poincaré's conventionalist philosophy. Poincaré referred to the theorem in (1897; 1954) and then again in 1900 (1902, 1952: pp. 167-8) to argue that there is always an excess of theoretical explanations over established facts and empirical-mathematical principles which summarise and generalise those facts:

> "Every time that the principles of least action and energy are satisfied... not only is there a mechanical explanation possible, but... there is an unlimited number of such explanations. By means of a well-known theorem due to Koenigs it may be shown that we can explain everything in an unlimited number of ways, by connections after the manner of Hertz, or, again, by central forces...."

Koenig's whole theoretical kinematics is, however, important as part of the background of Poincaré's conventionalist philosophy for another reason. Koenigs (who studied under Bouquet, Tannery and Darboux) treats kinematics as a geometry of displacements (hence time as an arbitrary auxiliary variable) from the point of view of transformation groups, i.e. as a study of the invariants with respect to various continuous groups. Theoretical kinematics so conceived is close to Poincaré's view of geometry as a study of the laws of displacements (Poincaré, 1887; 1898a) apart from the fact that he saw the origin of geometry in physical kinematics. Moreover, as was mentioned previously in connection with Klein's paper (1910) on the geometry of the Lorentz group, the mathematical tradition extending from Poncelet through Cayley to Lie, Klein, Poincaré, Darboux, etc. which emphasised the search for the invariants with respect (relative) to transformation groups, was presumably the source of the mathematical formulations of the principles of relativity of both the classical and the new mechanics.

From 1885 to 1896 Poincaré held the chair of mathematical physics at the University of Paris. In 1889 he published *Leçons sur la théorie mathématique de la lumière*. In 1888-1889 he gave lectures on optics and electricity, which resulted in *Electricité et optique: les théories de Maxwell et la théorie electromagnetique de la lumière* (1890). These lectures and the book based on them played an important role in popularising Maxwell's theory in France. Each of the mentioned books contains certain philosophical conclusions, concerning the epistemological status and role of theories, drawn from the survey of theories of mathematical physics given in it. Poincaré concludes his survey (1889: pp. 398–401) of the theories of light by saying that in the discussion of optical phenomena he exhibited several parallel theories all of which are able to account for the facts equally well. The theories in question could be classified into two groups: those which like Fresnel's assumed the elasticity of the medium to be constant and those which like Neumann's assumed the density of the ether to be constant. Although Fresnel's hypothesis was somewhat better in its explanation of aberration, none of the existing theories gave a satisfactory account of that phenomenon. Hence, there was no good reason for choosing one of the theories rather than another. However, we must not complain that it is impossible to make a choice, for this impossibility shows that mathematical theories of physical phenomena must not be considered as anything but *instruments of research* (Poincaré, 1889: p. 398). These instruments are very precious, indeed, but we must not be enslaved by them and ought to abandon them whenever they are contradicted by experimental findings (Poincaré, 1889: p. 399). There is, moreover, a general, formal reason for our inability to discriminate on experimental grounds between these two groups of optical theories. The equations of movements in an elastic isotropic or anisotropic medium are linear equations with constant coefficients. If ξ_1 and ξ_2 are two integrals of such an equation so is $A\xi_1 + B\xi_2$. In effect we have infinitely many ways of solving our problems in optics. Moreover, one of the equations of motion in an elastic, isotropic medium is as follows:

$$\frac{d^2\xi}{dt^2} = \Delta\xi - \frac{d\Theta}{dx} \qquad (I)$$

If a function satisfies equation (I) any of its derivatives will also do so. If we designate by ξ, η, ζ the components of the displacement of a molecule of the ether in Fresnel's theory, quantities which satisfy equations of the type (I), their derivatives ξ'_x, ξ'_y, ξ'_z, η'_x... also satisfy such equations, as do $(\zeta'_y - \eta'_z)$, $(\xi'_z - \zeta'_x)$, $(\eta'_x - \xi'_y)$. The latter, however, are exactly components of the displacement of a molecule of the ether in Neumann's theory, hence they satisfy Neumann's equations. One can be sure, therefore, that all phenomena explained by Fresnel's theory will also be explained by Neumann's and *vice versa* (Poincaré, 1889: pp. 399–400).

The epistemological point in Poincaré's comparison of Fresnel's theory with Neumann's appears to be this. In any branch of mathematical physics, optics for example, there are several, so to speak parallel, theories. They appear to be rivals but are observationally equivalent; whenever the solutions of the equations of one are given, these can be transformed into the solutions of the other.[19] One cannot choose between the theories on logical and experimental grounds and there is no need to choose, provided we regard them as *research instruments* and *do not assert* anything that goes beyond their common experimental and formal-mathematical content. Since Poincaré refers here to theories as merely research instruments, his epistemology has been often classified as *instrumentalist*. However, it should be noted that there is no trace here either of the thesis of ontological instrumentalism (phenomenalism) which denies the existence of unobservable objects, nor even of the thesis of semantical instrumentalism, which denies that theoretical terms have physical interpretations. Poincaré's "instrumentalism" amounts to the thesis of pragmatic instrumentalism (suspend judgement) combined with the somewhat sceptical doctrine that observationally equivalent theories which have similar mathematical structures form the limit of (theoretical) knowledge. In other words, Poincaré's "instrumentalism"—if that term is still applicable here—is equivalent with the philosophy of "the physics of the principles", viz. to systematise in terms of mathematical principles the experimental content common to "rival" theories and leave the choice of theories open.

In 1895 *L'Éclairage électrique* (April, May, October and November) published a large memoir by Poincaré entitled "A propos de la théorie de M. Larmor" (1895, 1954: pp. 369–426). This extensive survey of most theories in optics and electrodynamics of the time was to become the first of a series of such survey-articles (subsequent ones in 1900, 1904 and 1908 at a more popular level) which gave Poincaré the opportunity to summarise theoretical developments and main experimental results in the area, to make critical comments, propose mathematical improvements and expound his philosophy of physical theories. In the first instalment (April) concerned with the theories of Fresnel, Neumann, Maxwell and Larmor, the proximity (if not equivalence) of Poincaré's epistemology with that of the philosophy of the physics of the principles becomes particularly clear. Since this was one of the main bases for his more popular, philosophical paper "Sur les rapports de la physique expérimentale et de la physique mathématique" presented at the International Congress of Physics in Paris in 1900, made widely accessible in Chapters IX and X of *Science and Hypothesis* ("Hypotheses in physics" and "The theories of modern physics", respectively) (1902, 1952) and rightly seen as the epitome of Poincaré's conventionalist view of physical theories, it will be useful to recall some of its points. At the beginning Poincaré resumes his comparison of the two groups of theories in physical optics by introducing—without any hypothesis concerning their significance—two vectors, which proved to be of great importance in the field. In an isotropic medium (assumed for simplicity), the first of the vectors—which Poincaré names after Fresnel with components X, Y, Z,—is perpendicular to the plane of polarisation; the second vector L, M, N—Neumann's vector—is parallel to the plane of polarisation. If A stands for the reciprocal of the velocity of light c and ϵ denotes the square of the index of refraction, then the relations between the two vectors are expressed by equations:

$$\left. \begin{array}{l} A \dfrac{dL}{dt} = \dfrac{dZ}{dy} - \dfrac{dY}{dz}, \\ A\epsilon \dfrac{dX}{dt} = \dfrac{dM}{dz} - \dfrac{dN}{dy} \end{array} \right\} \quad (1)$$

and four other equations obtained by permuting letters x, y, z; X, Y, Z; L, M, N. Equations (1) which state that the derivative with respect to time of each of the two vectors is proportional to the *curl* of the other vector, "summarise, so to speak, the principal experimental facts of optics independently of any theory" (Poincaré 1895, 1954: p. 370). Different theoretical interpretations of the two vectors result in divergent theoretical claims.[20] Equations (1), therefore, play the same role *vis à vis* Fresnel's and Neumann's theories (or two groups of theories based on similar hypotheses) in physical optics which Hamilton's principle plays in geometrical optics *vis à vis* the particle and wave theory of light (cf. Chap. 2 in present volume). The idea that it is both scientifically and epistemologically instructive to extract in this way the common experimental and mathematical core of two apparently rival theories and that there is no need (for the time being at least) to commit oneself to the acceptance or rejection of either theory, is at the bottom of the philosophy of Hamilton and of Poincaré. There is, however, another common point, viz. the belief that our intellect is able to anticipate and guide experience by making genuine discoveries about the external world with the help of mathematical instruments, viz. by noticing formal or structural analogies. Such a striking formal analogy was noticed by Maxwell between the equations of optics and the experimental laws of the theories of electricity and led him to the discovery of the identity of light and electricity, subsequently confirmed by Hertz's experiment. The formal analogy in question is even more striking if one uses Hertz's notation (Hertz, 1890: pp. 575–624; 1894, 1956: pp. 195–240). If X, Y, Z now denote the components of the electric vector, L, M, N, the components of the magnetic vector, ϵ stands for the specific inductive capacity (dielectric constant) of the substance, then for non-magnets and non-conductors the mentioned quantities are related by equations identical with equations (1). For the magnetic field and conductors two more parameters are necessary: the coefficient of magnetic permeability (Magnetisirungskonstante) μ and the specific conductivity of the body λ. The equations then take the following form:

$$\left.\begin{aligned} A\mu \frac{dL}{dt} &= \frac{dZ}{dy} - \frac{dY}{dz}, \\ A\epsilon \frac{dX}{dt} &= \frac{dM}{dz} - \frac{dN}{dy} - 4\pi A\lambda X. \end{aligned}\right\} \quad (2)$$

From equations (2) one obtains equations (1) putting $\mu = 1$ and $\lambda = 0$, i.e. equations (1) are a special case of (2) (Poincaré, 1895, 1954: pp. 373–4).

In order to take into account matter in motion, equations (2) have to be replaced by more complicated ones. "A propos de la théorie de M. Larmor" (1895, 1954) contains also a survey of the theories of electrodynamics of Helmholtz, Hertz, Larmor, Lorentz and others. The least unsatisfactory of those Poincaré pronounced to be Lorentz's theory (1892), revised in (1895), based on the hypotheses of "charged particles" (called "ions" in 1895 and "electrons" from 1899 on) and of stationary ether. One of the consequences of electrodynamic theories of this type concerns the effects which the velocity through the ether should have on optical and other experiments carried out on moving bodies, the Earth, for example. If attempts to discover such effects proved positive, this would not only confirm the hypothesis of the ether but also justify the assignment of physical meaning to the concept of absolute velocity (or velocity relative to the stationary ether). However, Poincaré concludes that

... Experience has revealed a multitude of facts which can be summarised under the following formula: it is impossible to discover the absolute motion of matter, or still better the motion of ponderable matter relative to the ether. The proposed theories [which Poincaré surveyed—J.G.] explain this law fairly well on two conditions: Firstly, one must ignore dispersion and other secondary phenomena of the same type, secondly, one must ignore the square of aberration. However, this will not do; the law appears to be true even without those restrictions as has been shown in a recent experiment of Michelson" (Poincaré, 1895, 1954: p. 412).

The above passage contains the first formulation by Poincaré of the principle of relativity in terms of the impossibility of detecting absolute motion with the help of optical, electric, etc., effects even if quantities of second and possibly of higher order in v/c are taken into account. It is not surprising, therefore, that when in (1899; 1902, 1952: pp. 75–7) Poincaré formulated "the law of relativity" or "the principle of the relativity of space", to examine the relation between geometry and experimental physics, his law of relativity referred to *all laws* of physics and *not* to the laws of mechanics exclusively as does the classical principle of relativity (the same applies to "the principle of relative motion", already quoted from (1902, 1952: p. 111)). In other words, between 1895 and 1900 Poincaré generalised the classical and the electromagnetic principles of (experimental) relativity to one principle of relativity (relative motion) in effect equivalent to Einstein's special principle. When in the two papers on the dynamics of the electron (1905a, 1954) and (1906, 1954, 1970) he formulated the precise Lorentz transformations he identified the experimental version of the principle of relativity with the mathematical postulate of relativity according to which all laws of physics are form-invariant with respect to the Lorentz group. On that occasion he argued that if one assumes the principle of relativity as universally valid, then one finds that the velocity of light is a quantity common to the laws of electromagnetism, the law of gravitation and to all forces of whatever origin; this could be explained in one of two ways: either everything is of electromagnetic origin (as Lorentz believes) *or* else this is due to (the relational nature of) our methods of measurement (1906, 1954, 1970: p. 498).[21] Poincaré's principle of relativity given in the St Louis lecture, as indeed in 1900 and 1895, is equivalent to Einstein's special principle according to which "... the same laws of electrodynamics and optics will be valid for all frames of reference for which the equations of mechanics hold good..." (Einstein, 1905, 1970: p. 188). The passage quoted from Poincaré (1906) also shows that—contrary to Goldberg's and Miller's claim—Poincaré understood the light velocity to be constant with respect to all inertial frames and not with respect to the ether, otherwise the reference to the principle of relativity would not make sense. This can also be seen from his investigation of the invariants of the Lorentz group in (1906, 1954, 1970).

The epistemological comments concerning the relation between various theories in physical optics (typified by Fresnel's and Neumann's theories) and between Fresnel's and Maxwell's theory which we find in (Poincaré, 1889; 1895, 1954) are repeated in more elaborate form in (Poincaré, 1900, 1902, 1952: pp. 160–1) and in (Poincaré, 1904, 1905, 1958: pp. 139). In fact the latter become much more intelligible if we read them in conjunction with the analysis of Fresnel's, Neumann's and Maxwell's equations given in those earlier works of Poincaré. Such a confrontation clearly shows several important points.

Firstly, at least between 1889 and 1905 Poincaré's main epistemological concern, arising from his research and lectures on mathematical physics, was the defence of the ideal of the objectivity and rationality of scientific knowledge in the face of the proliferation and frequent change of theories in physics. In the 1900 paper (1902, 1952: p. 160) he speaks of the "ephemeral nature of scientific theories" as the source of many people's belief in the "bankruptcy of science" and argues that this scepticism is not justified if only one takes proper account of "the object of scientific theories and the part they play". In the 1904 lecture (1905, 1958: p. 139) he writes that "At the first blush it seems to us that the theories last only one day and that ruins upon ruins accumulate". However, if one looks more closely, "there is in them something which usually survives".

Secondly, Poincaré's defence of the objectivity and progress in science is based on his view of the relational nature of all knowledge. The point of theories is to provide numerical predictions and to express in the form of mathematical structures the relationships between apparently remote areas ("profound reality"). "Fresnel's object was not to know whether there really is an ether... his object was to predict phenomena...." Although Fresnel attributed light to the movements of the ether while according to the electromagnetic theory light is a current, not only predictions from Fresnel's theory survive but also his equations remain true and as such became absorbed into Maxwell's equations. Those equations teach us now, as they did then, that "... there is such and such a relation between this thing and that; only, the something which we then called 'motion', we now call 'electric current', but these are merely names for the images we substitute for the real objects which Nature will hide for ever from our eyes..." (Poincaré, 1902, 1952: p. 161; 1905, 1958: p. 139). Throughout frequent changes in physical theories what is being replaced are often only images, metaphors or "indifferent hypotheses",[22] i.e. conventions of a sort adopted because for the time being they appear as convenient aids to our mind or imagination. Naturally, not all theoretical changes are of this type: "... many connections that were believed well established have been abandoned". But "... there is something that remains over and this something is the essential. This it is which explains how we see the present physicists pass without any embarrassment from the language of Fresnel to that of Maxwell..." (Poincaré, 1905, 1958: p. 139).

Thirdly, Poincaré's view of the nature and aim of physical theory— which, as the above quotations show he held at least between 1889 and 1905 (i.e. the year he produced his mathematical paper on the dynamics of the electron with the mathematics of special relativity)— makes it inconceivable that he should see Lorentz's electron theory differently, viz. in a naïve-realist fashion and subscribe to Lorentz's electromagnetic research programme with all its metaphysics of electrons and the ether.[23] It is not coincidental, surely, that Poincaré resumes the question of the existence of the ether invariably in the context of his critical analyses of the nature and aim of the theories of mathematical physics or of the principle of relativity (negative results of attempts to discover absolute motion). (Poincaré, 1895, 1954: pp. 412–3; 1900, 1902, 1952: pp. 168–72; 1904, 1905, 1958: pp. 98–9, 139–40). The former would allow the ether only relational-mathematical properties (field-equations), the latter denies it the role of a preferred frame; neither could be reconciled with Lorentz's unexpurgated theory. In fact Poincaré's comments such as the one concerning the role of light in our measuring procedures (1906) or the other (previously quoted) concerning the highly hypothetical nature of contraction (realistically understood) amount to a re-interpretation of Lorentz's theory.

Conclusions

In 1895, 1900, 1902 and 1904 Poincaré formulated the principle of relativity in terms of the impossibility of detecting by any physical methods the absolute velocity of the earth or in terms of the invariance of physical laws in all inertial frames.

In 1897, 1898, 1902 and 1904 Poincaré outlined a programme for a new mechanics with relativistic features and predicted its realisation. Among those features were: the abandonment of "absolute space" and "time", of the ether hypothesis, the adoption of the principle of relativity as one of the supreme postulates, of light velocity as limiting, free choice of whichever geometry would prove more convenient. Since Einstein read Poincaré (1902), it is possible that the prediction turned out to be self-realising in that it prompted and encouraged Einstein to set up a new theory along these lines.

In 1902 and again in 1912 (1913, 1920, 1963: p. 21) Poincaré formulated the idea of what later came to be known as general covariance of laws.

In 1905 (1906, 1954, 1970), i.e. simultaneously with Einstein's (1905)—though the two papers were independent of one another—Poincaré produced most of the mathematics of "special relativity", viz. precise Lorentz transformations, the postulate of relativity in terms of the form-invariance of laws with respect to Lorentz transformations, the group properties of Lorentz transformations (the invariants of the Lorentz group), the covariance of Maxwell's and Lorentz's equations with respect to the Lorentz group, the elements of the four-vector formalism (in the context of a Lorentz invariant gravitation theory), together with the idea of Lorentz transformations as rotations in 4-space. So far as these components of special relativity are concerned, one can only speak of simultaneous discovery by Einstein and Poincaré.

In his 1906 memoir and subsequently (1908, 1956) Poincaré indicated that Lorentz's transformations may be given not only the physical interpretation attached to them by Lorentz (in terms of the Lorentz–Fitzgerald contraction, local time, etc.) but also in terms of measuring procedures for which the properties of light (in particular its finite and constant velocity) are essential. Although the latter interpretation is preferable from the point of view of Poincaré's conception of the content of a theory of mathematical physics (i.e. the predictive content and the type of mathematical structure, in this case the Lorentz group), neither his mathematics nor his philosophy excluded the former. If our concept of a theoretical discovery requires that the discoverer of a theory should be firmly committed to one unique physical interpretation, to the exclusion of all others, then—in this sense—Poincaré did not discover special relativity and could not have done so as long as he believed in the philosophy implicit in the physics of the principles and absorbed into the conventionalist epistemology.[24] For, on the one hand, the sceptical principle of that philosophy does not permit dogmatic commitment to *any* theoretical interpretation understood in the naïve-realist sense; on the other hand, the conventionalist principle of theoretical tolerance demands that one should respect a scientist's wish to *entertain* or work with any interesting theoretical interpretation of experimentally adequate principles. It would be a strange concept of theoretical discovery which would make it impossible—by definition—for any scientist holding the conventionalist view of theories in the sense explained to make any theoretical discovery.

Notes

1. In the debate claims have been made alleging, for example, Poincaré's intellectual conservatism and opposition to relativity theory and to Einstein personally. However, Poincaré's letter written in 1911 in support of the recommendation to appoint Einstein to a professorship at Zürich Polytechnic (a similar supporting letter having been written by Mme Curie) shows his great appreciation of Einstein's intellect (Hoffmann, 1975: p. 99) and his two articles on quantum theory written in the same year indicate how alert and open to new ideas was his own mind shortly before his illness and death in 1912. The circumstance that Poincaré never refers (with one or two small exceptions) in his writings to Einstein's contributions to relativity—which is presumably the basis for these allegations—is understandable in the light of Einstein's failure to make any acknowledgements in his relativity paper (1905) and in the light of Poincaré's belief that the same theory was already present in Lorentz's and his own works.
2. See Chap. 2 in this volume for a discussion of the concept of cognitive content from Poincaré's point of view.
3. "Imprimé le 9 mars 1915" but published with delay, caused by the war, in *Acta Mathematica*, **38** (1921).
4. Lorentz established, however, the invariance of the equations for charge-free space.
5. As a statement of chronological order this is correct only if Lorentz meant (Poincaré, 1905a) i.e. the short note "Sur la dynamique de l'electron" in *C. R. Hebd. Séanc. Acad. Sci.* rather than (Poincaré, 1906).
6. The inhomogeneous Lorentz group, i.e. the group of rotations and translations in the 4-space, is now usually called "the Poincaré group".
7. The term "approximate Lorentz-transformations" is avoided here since it has been used to refer to the transformation equations which appear in Lorentz's pre-1904 works (and in (Larmor, 1900)) and in which second order effects are disregarded.
8. In this context one would have to explain why there were no new references to Poincaré's contributions to relativity in the second, 1915, edition of *The Theory of*

Electrons. One plausible answer would be that Lorentz did not want to initiate a controversy over priorities, the more so that Poincaré was dead already and had not engaged in a controversy on the subject while alive. An *hommage* to Poincaré seemed, from that point of view, the best form to give credit where it was due without offending the other, innocent, party.

9. On 9 December, 1919, Einstein wrote to Born: "... Schlick has a good head on him; we must try to get him a professorship..." (Born, 1971: 18). In (1921, 1959: p. 34) Hans Reichenbach writes: "The first philosopher to accept the theory of relativity wholeheartedly was Schlick. In this regard, he occupies a leading position among philosophers, and his conception of the theory of relativity, which is related to Poincaré's conventionalism, is shared by Einstein. Schlick opposes Kant and represents philosophical empiricism, but he also defines his position in contrast to Mach's positivism."

10. In my (1973) "Logical comparability and conceptual disparity between Newtonian and relativistic mechanics"—Appendix in present volume—I argue that, following Philipp Frank's analysis, one can see the conceptual disparity between Newtonian and relativistic mechanics as due to their observational comparability, viz. as due to the fact that certain empirical laws tacitly assumed by Newton or Newtonians to ensure the uniqueness of the definitions of certain mechanical concepts are contradicted by the assumptions of Einstein's special relativity. In Schlick's analysis the conceptual disparity between Lorentz's and Einstein's theories is due to the logical comparability (contradiction) between the ontologies assumed from the outset by Lorentz and Einstein respectively. Once the theories have been constructed their respective ontological claims may be seen relative to the relevant theory. In a historical account of the genesis of each theory, its existential claims may have to be seen as absolute (in the sense of Quine's *Ontological relativity*, 1969, Chap. 2).

11. The reference de Sitter makes to Poincaré's (1906) memoir, viz. "Poincaré: 'Sur la dynamique de l'electron', *Rc. Circ. Mat. Palermo*, Vol. XXI, p. 129 (Dec. 1905)" suggests that it may have been a subtle contribution to the priority dispute. However, presumably de Sitter did not claim that Poincaré's memoir *appeared* in December 1905 in t. 21 of the *Rc. Circ. Mat. Palermo*. But the suggestion that it was *submitted* in December 1905 would be incorrect since, in fact, the *Rc. Circ. Mat. Palermo* memoir was dated "Paris, juillet 1905" and the *C. R. Hebd. Séanc. Acad. Sci.* abstract "5 juin 1905".

12. Plummer (1910) is based mainly on Lorentz (1904) and gives to Lorentz the credit for the principle of relativity. In Section 8 the new laws of aberration and of the Doppler effect are discussed as derived by Einstein, with a due reference to his (1905). In the last Section, 14, reference is made to Poincaré's application of the principle of relativity to gravitation theory in (Poincaré, 1905a). Plummer (1910) and Whittaker (1910) have been brought to my attention by Professor W. H. McCrea whose bibliography of E. Cunningham's works also proved very helpful.

13. The subject of Klein's paper (1910), presented on 10 May to the Göttingen Math. Society, had been discussed by him with Hermann Minkowski, just before the latter's death. In an outline of the history of Lorentz transformations in his 1915–17 lectures (Klein, 1927, 1956: pp. 70–5), Klein mentions the contributions of: W. Voigt, 1887; Lorentz, 1892; Larmor, 1900; Poincaré, 1905, 1958; Einstein, 1905; and Minkowski, 1907, 1908. Concerning Poincaré's and Einstein's contributions Klein writes:

"... Now comes the year 1905 with its decisive publications by Poincaré and Einstein who proceed independently side by side, although both speak of the 'Postulate' or the 'Principle' of 'Relativity'. Poincaré makes the beginning with a note 'Sur la dynamique de l'electron' in the *C. R. Hebd. Séanc. Acad. Sci.* of the Paris Academy on June 5th, which was then submitted as a longer memoir to the Circolo Matematico di Palermo on June 23rd [rather, July—J.G.] but as such appeared as late as 1906... Einstein's work 'Zur Elektrodynamik bewegter Körper', on the other hand, was submitted to the editor of the *Annalen der Physik* on June 30th and appeared already on September 26th in vol. 17. It is very interesting to compare such competing publications—Poincaré presents the mathematical tools more clearly.... In Einstein's work, on the other hand, the nature-philosophical thinking is in the foreground..." (Klein, 1927, 1956: p. 73). "Competing publications" ("konkurrierende Veröffentlichungen") in my understanding does *not* concern here the content of the two papers, i.e. the theories they contain, but rather the simultaneity of their publication.

14. It should be pointed out that the doctrine of the theory-dependence of the meaning of all scientific terms and the associated "incommensurability thesis", so much discussed in contemporary philosophy of science, had been anticipated by conventionalists, with approval by the extreme ones (LeRoy, Duhem, Ajdukiewicz), with disapproval by others (Poincaré). Theories which are not intertranslatable, i.e. "incommensurable" ones, behave in certain important respects exactly as do observationally equivalent theories. Both types are obviously insensitive to any crucial experiments and, therefore, can be seen as rivals only in some sense not envisaged in either the apriorist or the traditional empiricist rationality. This explains why the thesis that "logic and experiment alone" are insufficient to account for the choice of actual scientific theories, supposed to be characteristic of the so-called "new empiricism" of today, had been the hallmark of the extreme and moderate forms of conventionalism. Conventionalism in general was a reaction to (and a partial continuation of) the apriorist and traditional empiricist rationality: whereas its extreme (nominalist) variety emphasised the possibility of total meaning—change with the change of conventions, the moderate—Poincaréan—variety emphasised the existence of observational and formal-structural "invariants" and the double role of (observationally and formally) equivalent theories: their roles in providing comparability and progress in the flux of theoretical-metaphorical and indifferent-hypothetical conventions, as well as their role in defining the limits of theoretical knowledge.

15. In fact, the phrase quoted by Holton had been used some eleven years before by Hans Reichenbach in "The philosophical significance of relativity theory" (Reichenbach's contribution to 1949, 1969: *Albert Einstein* edited by P. SCHILPP, p. 292) to make the point that a creative physicist, like Einstein, may and even has to be uncritical about the philosophy that stimulates his physics, but a philosopher of science, who subsequently examines physical theories and the philosophy associated with them has to apply critical standards. In general, however, the distinction between the context of discovery and the context of justification was used by logical empiricists for two purposes: firstly, to effect a division of labour between the history and the philosophy of science, secondly, to avoid various kinds of the "genetic fallacy", in particular, the early (genetic) empiricists' claim that the status and justification of a hypothesis or theory depend on the way they have been discovered or arrived at, e.g. whether by the generalisation of observed

facts or otherwise. As we have seen, Schlick, the co-founder of the Vienna Circle, used "the context of discovery" to establish the conceptual disparity between Lorentz's and Einstein's theories, their observational equivalence notwithstanding.

16. This is a revival and rehabilitation of the hermeneutic conception of "meaning" or "sense", which originated in the nineteenth century, mainly German, tradition of the "Geisteswissenschaften" and "Lebensphilosophie" and was reborn—at first oblivious of its provenance—in the second philosophy of Wittgenstein as well as in the writings of many of his followers. What the logical-empiricist critics of the nineteenth-century hermeneuticists saw as a confusion of the concept of the meaning (or sense) of linguistic expressions and the "meaning" (or "sense"), i.e. aim, purpose, function, value, etc., of human actions, artefacts, of life, etc., has been elevated to a virtue, to a principle of semantics and philosophy. The same applies to the concepts of "value-relatedness" and "subjectivity" (or "impossibility of objectivity"). Each of these (con-)fusions is associated with one of the two main operations of the hermeneutic mind: empathy (sympathetic, humanistic, etc., understanding) and unmasking (demystification). Many of the issues involved here have been discussed in Hesse (1980).

17. Holton applies this unmasking approach for example in his criticism of Whittaker's repeated reference to Lorentz's (1904) memoir as if it had appeared in 1903. In Holton's words:
"... It is more difficult to discuss the 1903 paper of Lorentz which Whittaker, both in his book and in his Memoir, cited specifically as the work that spelled out most of the details of Einstein's Relativity Theory of 1905. In the first place, this paper does not exist. What Whittaker clearly wished to refer to is the paper Lorentz published a year later, in 1904. Since Whittaker was otherwise very careful with voluminous citations of references, this repeated slip, which doubles the time interval between the work of Lorentz and Einstein is not merely a mistake. It is at least a symbolic mistake, symbolic of the way a biographer's preconceptions interact with his material" (Holton, 1960, 1974: p. 177).

This point in Holton's paper has been criticised already by Keswani (1965: p. 287, footnote 3) who pointed out correctly that Whittaker gives 1904 as the year of publication of Lorentz's memoir on other pages in his book. Keswani concludes "... obviously, mention of the year 1903 elsewhere was only a mistake, pure and simple". Having read these exchanges I was puzzled when one day I found in E. Cunningham (1910: "The principle of relativity in electrodynamics and an extension thereof", *Proc. Lond. Math. Soc.* 2, 2, 77) the following reference to Lorentz's memoir in question: "Lorentz, Amsterdam Proceedings, 1903–4, p. 809". Though roughly Whittaker's contemporary and also a Cambridge mathematician, Cunningham in his articles and books on relativity (one of which was, according to W. H. McCrea, the first treatise on the subject in English) was clearly on Einstein's side, as regards the priority issue, and concerned that due credit should be given to J. Larmor, which was unaffected one way or the other by the question whether Lorentz's memoir appeared in 1903 or in 1904. An inspection of Volume VI of the *Proc. Sect. Sci. K. Ned. Acad. Wet.* resolved the mystery. Volume VI consists of two parts, the first of which appeared in December 1903 whereas the second appeared in July 1904. Lorentz's memoir is in Part 2 but the correct reference to Volume VI of the *Proceedings* is 1903/4, as in Cunningham. This is, in all probability, the source of Whittaker's inconsistent references.

18. One has the impression that Poincaré's critical analysis of the conventions of absolute space and time assumed in classical mechanics as well as his discussion of the relativity of position, space, time and motion, were written with Euler's "Réflexions sur l'espace et le temps" (1748) in mind. Euler argues for the reality of absolute space and time (against the metaphysicians' claim that these are mere fictions), from the premises that, firstly, the laws of Newtonian mechanics (e.g. of inertia) require the concepts of absolute space and time and that, secondly, the laws of mechanics could not be incontestable, indubitable truths which they are if they were based on fictions.
19. Hence, there is a one-to-one correspondence between any model of the observational part (the Ramsey-sentence formulation) of Fresnel's theory and a model of the observational part (Ramsey-sentence) of Neumann's theory.
20. In other words, models of the observational part of Fresnel's and Neumann's theories, which are observationally equivalent, are extendible to the full models of either theory.
21. In his St. Louis lecture (1904, 1905, 1958: p. 94) Poincaré formulated again the principle of relativity according to which "the laws of physical phenomena must be the same for a stationary observer as for an observer carried along in uniform motion of translation; so that we have not and cannot have any means of discovering whether or not we are carried along in such a motion". This is one of the passages to which Whittaker refers in his (1953, 1973) when he argues for Poincaré's priority. Holton's claim that the quoted statement expresses merely the classical Newton–Galileo principle of relativity seems to me untenable, if by "classical" one means the principle to the effect that it is impossible to distinguish between inertial frames by any *mechanical* method. Not only is there no such restriction in Poincaré's own wording but the context would not permit this interpretation. The point Poincaré is arguing in the 1904 lecture is that although the principle of relativity had been assailed by numerous experimental attempts to discover the effect of the motion of the Earth through the ether on optical and electric phenomena predicted by various optical, electrodynamic and astronomical theories based on the ether hypothesis, nevertheless it has survived intact in view of the negative results of all these attempts. "Michelson has shown us, I have told you, that the physical procedures are powerless to put in evidence absolute motion; I am persuaded that the same will be true of the astronomic procedures, however far precision be carried" (Poincaré, 1904, 1905, 1958: p. 100).
22. "In optical theories two vectors are introduced, one of which we consider as a velocity and the other as a vortex. This again is an indifferent hypothesis, since we should have arrived at the same conclusions by assuming the former to be a vortex and the latter to be a velocity. The success of the experiment cannot prove, therefore, that the first vector is really a velocity. It only proves one thing—namely, that it is a vector; and that is the only hypothesis that has really been introduced into the premises. To give it the concrete appearance that the fallibility of our minds demands, it was necessary to consider it either as a velocity or as a vortex..." (1902, 1952: pp. 152–3).
23. In (1900a; 1954: p. 464) Poincaré wrote with reference to Lorentz's theory: "... all good theories are flexible.... If a theory reveals some true relations, it can assume a thousand different forms, it resists all assaults and what is essential in it, does not change..." and "... good theories... triumph over objections, even those which are serious, but they triumph by transforming themselves...."

24. In 1904, as we have said. Poincaré distinguished three periods in the evolution of modern physics, viz. the physics of central forces, the physics of the principles and the third period around the turn of the century. Since his own conventionalist epistemology was largely—as regards physics—an elaboration of the philosophy implicit in the physics of the principles, one might think that perhaps it was no longer descriptively adequate of the third period, which in many ways was a revival of the tendencies of the first period (search after theoretical explanations in terms of ultimate constituents of the universe) and that Poincaré would see this. There are indications that he did. However, what conventionalism took from the physics of the principles was not phenomenalism but rather the tolerance for a variety of theoretical interpretations of experimentally adequate principles. It was this tolerance combined with the suspension of judgement and with the idea that mathematical structures (or types of mathematical structures) have also a descriptive-cognitive function, that made conventionalism an attractive philosophy in the defence of the traditional values of objectivity and progress in science.

A recent interesting attempt to define kinds of mathematical structures which might be used to characterise different physical theories may be found in Scheibe (1979).

Appendix

Logical Comparability and Conceptual Disparity Between Newtonian and Relativistic Mechanics*

ARTICLES and notes published over the last ten years or so and dealing with the question of so-called incommensurable theories in science have resulted from the controversy over the structure, changes and rational evaluation of scientific theories. Under attack have been a number of philosophical views of science: from logical empiricism, mainly in the works of Carnap, Reichenbach and Hempel, through a variety of empiricist doctrines propounded by professional philosophers, e.g. Braithwaite and Nagel, or by some physicists, e.g. Bridgman, Bohr, Einstein, to earlier critics of logical empiricism, e.g. Popper. The attack was originally launched by Hanson, Toulmin, Kuhn and Feyerabend, but many others joined in.

Some criticisms were meant to apply exclusively to the "logical approach" identified with logical empiricism. So, for example, it was suspected that

> ...the view that scientific theories are interpreted axiomatic systems may have blinded its adherents to many of the functions of those theories and their components... even the highly developed scientific theories on which the axiomatic approach concentrates may be inadequately treated when looked upon as mere interpreted axiomatic systems. For the logician deals with those theories and their constituents as static, frozen in a logical mould; but perhaps there are more "dynamic" functions such an approach tends to make us overlook... (Shapere, 1965: p. 28).

One criticism, however, was meant to apply not only to logical empiricism but to all those mentioned as being under attack: it is, that they have failed to see the existence and importance in science of rival, incommensurable (logically and empirically non-comparable) theories and that—owing to that failure—they have over-rationalised their philosophic accounts of science, seeing or demanding logical relations (consistency, incompatibility, deducibility, reducibility, definability, etc.) where there are none. The critics seem to regard as a new and important insight and as their contribution to the philosophy of science the claim that so-called revolutionary changes in science are disruptive, since they bring about radical changes in fundamental scientific concepts owing to which either piecemeal or even wholesale logical comparisons of rival theories may be impossible; this claim sets, according to critics, limits to rational (here: logical and empirical) arguments in science.

In two previously published articles (Giedymin, 1968; 1970) I pointed out that the so-called "incommensurability thesis" both in its general form, i.e. as the claim that there are incommensurable theories in science, and in its specific applications, e.g. the claim that Newtonian and relativistic mechanics are such incommensurable theories, is certainly *not* a new insight, since it was explicitly formulated and systematically discussed by the author of radical conventionalism, Ajdukiewicz, in three articles published in the logical empiricist *Erkenntnis* in the early nineteen-thirties. For completeness sake I should like to add now that the general thesis dates back to the turn of the century when it was formulated and discussed in the controversy between Poincaré and LeRoy (Poincaré, 1905, 1958), the latter arguing in its favour and the former against. Ajdukiewicz (1934) resuscitated LeRoy's thesis, based it on a more precise (pragmatic) conception of language and meaning, mentioned Newtonian and relativistic mechanics as instances of not intertranslatable languages and finally, in 1936, abandoned the thesis of radical conventionalism as too extreme and untenable.[1] It should be obvious, therefore, that logical empiricists—thanks to the influence exerted on the philosophy of modern empiricism by conventionalists—were not only familiar for quite a long time with the idea of "incommensurability" (non-translatability, conceptual disparity) and disruptive changes in science but apparently came to the conclusion that the radical incommensurability thesis was untenable.

The present essay is intended to draw the reader's attention to the analysis of the relation between Newtonian mechanics and special relativity mechanics, given by Philipp Frank, one of the classics of logical empiricism (Frank, 1946, 1971). Frank's analysis of the relation between Newtonian mechanics and special relativity mechanics is interesting in many ways. Firstly, it shows clearly that problems of disruptive changes and of conceptual disparity were known to and discussed by logical empiricists and that, therefore, some of them at least were not at all blinded to dynamic problems of changes in actual science by their view of physical theories as interpreted axiomatic systems[2] and did not deal with those theories as "static, frozen in a logical mould". Secondly, it reveals some of the reasons for the logical empiricist rejection of what came to be known later as the incommensurability thesis.

Frank, just as Ajdukiewicz and many others before him, was aware of the conceptual disparity between Newtonian mechanics and special relativity mechanics. He discussed in detail the syntactical and semantical differences between the fundamental concepts of the two theories ("mass", "time distance", "length", "force") on which claims of indefinability, impossibility of translation, etc., between the two theories have been based, On the other hand, *on Frank's account Newtonian mechanics and special relativity mechanics are mutually inconsistent and, therefore, logically comparable* in spite of conceptual disparity; or—more interestingly—*Frank attempted to show that partial conceptual disparity may result from logical incompatibility of two rival theories*, without making the apparently inconsistent claim that Newtonian mechanics and special relativity mechanics are both incompatible and incommensurable. I am going to suggest in the present essay that Frank's account of the relation between Newtonian mechanics and special relativity mechanics can be generalised, under certain assumptions, to similar cases of rival theories using Carnap's ideas of indirect interpretation of theoretical terms and of the meaning postulates formulated in terms of the Ramsey sentence $^R T$ of a theory T. This will show, I hope, how certain conclusions drawn from case studies of actual scientific theories and general logical considerations fit together within the logical empiricist account of science.

Frank's analysis of the relation between Newtonian mechanics and special relativity mechanics is, in outline, as follows.

Newtonian mechanics and special relativity mechanics are mutually inconsistent, therefore, logically comparable. For from special relativity mechanics one can deduce the negations of certain empirical laws which have to be assumed as valid in Newtonian mechanics in order to ensure the uniqueness of the operational definitions of the fundamental concepts of Newtonian mechanics, "mass", "time interval", "distance", "force". If special relativity mechanics is true (or, if special relativity mechanics is assumed hypothetically as true), then those empirical laws, assumed in Newtonian mechanics to define the mentioned concepts, are false (or, have to be assumed false). Consequently, *Newtonian definitions no longer satisfy the condition of uniqueness, the concepts in question "have no operational meanings"*, as Frank, following Bridgman, put it, or—to use different words—are vague, i.e. have varying denotations (extensions) (cf. Przełęcki, 1969: pp. 18–22, 28). To become empirically meaningful from the point of view of special relativity mechanics, they have to be re-defined. On the other hand, if Newtonian mechanics is true (or, if Newtonian mechanics is hypothetically assumed as true), then the concepts of Newtonian mechanics do have empirical meanings, i.e. they have unambiguously fixed physical denotations. It follows that *if special relativity mechanics is true* (or is assumed to be so), *then the Newtonian concepts cannot have the same interpretations* (in the extensional sense) *which they are supposed to have on the assumption that Newtonian mechanics is true*. To conclude, according to Frank, there is *conceptual disparity between Newtonian mechanics and special relativity mechanics* which, however, far from making the two theories logically and empirically "incommensurable", is *due to the mutual inconsistency* (i.e. logical comparability) *between the two theories*.

Let us recall some of the examples of such conceptual changes discussed by Frank. Consider first the definition of "one hour" in terms of "the time during which the big hand of our pocket watch traverses an angle of 360 degrees". Obviously, we mean that any pocket watch can be used and not necessarily ours. However, this amounts to assuming that "the hands of all pocket watches proceed with one and the same angular velocity", which is a statement of a physical law about the behaviour of watch springs. Moreover, we mean any clock to be used, e.g. a pendulum clock and this, in turn, amounts to defining "one hour" as "a duration of a certain number of oscillations of a pendulum". These, apparently different definitions are equivalent if the following law is valid. "The unwinding of a spring as an effect of its elasticity proceeds at a rate which is proportional to the frequency of the pendulum as an effect of gravity." (Frank, 1946, 1971: p. 432) assumed in Newtonian mechanics. Now, from the postulates of special relativity mechanics it follows that a clock which travels with the speed v relative to S loses time compared with a clock at rest in S. Consequently:

> ... some operations which rendered, according to Newton's laws, identical results no longer do so if Einstein's principles are assumed to be true. The operations by which the time distance between two events was defined did not mention the speed of the clock relative to any system of reference. For, according to Newton's physics, this speed is without influence upon the march of the clock. If Einstein's principles are true, this operational definition of the time distance between two events becomes ambiguous. We must specify the speed of the clocks used in measurement. In order to obtain an unambiguous result of our defining operation, we must no longer say that "between the events A and B there is time distance of 10 seconds" but that "there is a time distance of 10 seconds if we use clocks which are at rest in a particular system $\bar{S}$." The velocity of $\bar{S}$ relative to S must be specifically given. We use a "relativised language" in order to make the description and the operations unambiguous (Frank, 1946, 1971: p.456).

Again, consider Newton's second law as a definition of "force" and then of "mass". To make the formula "$f = ma$" an unambiguous definition of "f" "... certain physical effects must be confirmed", namely "... the product ma has to be independent of the mass m and has to depend on the situation of the moving body in its environment..." (Frank, 1946, 1971: p. 441). Alternatively, assuming that the field of force is known, e.g. given by Coulomb's law, one gets the mass m by measuring the acceleration a:

> ... According to Newton's mechanics the result is independent of whether the initial velocity was small or great. But if Einstein's principles are right, this operational definition becomes ambiguous. The acceleration a (and, therefore, m) depends actually upon from what initial velocity we start the experiment. In order to obtain an unambiguous result, we have to specify the operation involved,

in particular the initial velocity v. If we require that the initial velocity be zero relative to (the fundamental system) S, the acceleration becomes unambiguously determined. We must, therefore, use a modified operational definition of "mass". We can either make the specification that the initial velocity relative to S is zero, then we define a concept which is called "rest mass" m_0. Or we can include the initial velocity v in the description of the operation. Then acceleration and mass themselves become dependent on v. We obtain a physical quantity which is no longer a constant but a function of v. This quantity is called "mass" in the new mechanics. By using this definition we can formulate the laws of motion in the simple form: mass times acceleration equals force ($ma = f$). But the mass m is now a function of v (Frank, 1946, 1971: pp. 455–6).

I shall now recapitulate some of the main points of Frank's analysis and make a few comments.

1. On Frank's account *Newtonian mechanics and special relativity mechanics are mutually inconsistent* and, therefore, logically comparable. In spite of this, or rather owing to this fact, there is conceptual disparity between the two theories, i.e. the specific, theoretical concepts of Newtonian mechanics have undergone changes in the transition to special relativity mechanics. This claim, that Newtonian mechanics and special relativity mechanics are logically comparable and yet to some extent conceptually disparate, is—of course—not peculiar to Frank's viewpoint.

2. The mechanism of conceptual change in the transition from Newtonian mechanics to special relativity mechanics is, according to Frank, as follows.

Some empirical laws have to be valid to ensure the uniqueness of the definitions of metric terms. If those laws happen to be false, the terms in question become vague and lose their empirical ("operational") meaning. If a theory T' logically implies the negation of the laws assumed in another theory T to define the terms of T, then from the viewpoint of T' those definitions have to be corrected (e.g. relativised) and so the terms in question have to be reinterpreted.

Special relativity mechanics implies the negation of certain empirical laws assumed in Newtonian mechanics to ensure the uniqueness of the definitions of "mass", "distance", "time interval", etc. So, for example, the sentence "A clock which travels with the speed v relative to S loses time compared with the clock at rest in S" implies the negation of the sentence "The speed of the clock relative to S is without influence upon the march of the clock"; the former sentence is a consequence of special relativity mechanics and the latter is implicitly assumed in Newtonian mechanics.

The empirical meaninglessness of Newtonian concepts from the viewpoint of special relativity mechanics is the result of factual considerations, i.e. of the denial of certain empirical laws assumed in Newtonian mechanics.[3]

To remove the vagueness of Newtonian concepts one has to take into account certain factors disregarded in Newtonian mechanics and this will affect also the syntax of the language, e.g. Newtonian "mass" is an expression of the form "$m(x) = y$", whereas relativistic "mass" is an expression of the form "$m(x, v) = y$".

3. The examples discussed by Frank fall under a schema and can be generalised using Carnap's ideas of "reduction sentence", "meaning postulate", and "the Ramsey-sentence of T".

If a term 't' is introduced into a theory T with the help of one reduction sentence of the form:

$$\text{If } 0_1(x), \text{ then } (t(x) \text{ iff } 0_2(x))$$

then "t" remains uninterpreted (i.e. has no empirical meaning) whenever $0_1(x)$ is not satisfied. Similarly, if an observational consequence of the reduction sentence or of a pair of reduction sentences introducing "t" turns out to be false, then again "t" remains empirically uninterpreted (Carnap, 1936; 1952). Now, definitions of metric terms such as "mass", "distance", "time interval", etc., are essentially of the same form as the above reduction sentence, except that the definiendum is an expression of the form "$m(x) = y$" or "$m(x, v) = y$", etc., while the definiens specifies the measuring operation, the measuring apparatus and its behaviour (e.g. pointer-readings).

In general, if $^R T$ is the Ramsey-sentence of a theory T, expressing T's observational content, then the conditional "$^R T \to T$" is the analytic component or meaning postulate of T, intended to provide the theoretical terms of T with empirical interpretations (in the extensional sense). If, therefore, $^R T$ happens to be false, the theoretical terms of T remain completely vague, i.e. have no empirical interpretation. Now, we may have grounds for accepting the negation of $^R T$ either on the basis of direct falsification of some of its consequences or else on the basis of accepting another theory T' which logically implies the negation of $^R T$. The latter would presumably be the case of Newtonian mechanics and special relativity mechanics on Frank's account, with the already mentioned consequence that from the point of view of special relativity mechanics the theoretical terms of Newtonian mechanics become empirically uninterpreted (we identify here Frank's "has no operational meaning" with "is empirically uninterpreted" or "is completely vague").

Needless to say, at the time when Frank's article appeared (in 1946) only the older version of Carnap's reconstruction of empirical theories (Carnap, 1936/7; 1937) was known in which two components of theories were distinguished, viz. the uninterpreted calculus (specific to the given theory, e.g. Maxwell's equations of the electromagnetic field, as well as the mathematics and logic necessary for making deductions) and "semantical rules" identified by Frank with "operational definitions" for metric terms.

4. Frank's claims that Newtonian mechanics and special relativity mechanics are mutually inconsistent and that some conceptual changes in the transition from the one to the other were due to this inconsistency are based on an intuitive formulation (one of several possible) of both theories and on the intuitive concepts of deducibility and inconsistency. Presumably Frank believed that in spite of partial conceptual disparity the languages of the two theories do have some sentences in common; sentences concerning the behaviour of measuring instruments (rods, scales, clocks, etc.), i.e. pointer-readings, were classified by him as common to both theories. It is plausible, however, to interpret him as saying that also some of the more theoretical sentences, which depend on background theories, are shared by Newtonian mechanics and special relativity mechanics. Following Einstein[4] and others, Frank discussed the relation between Newtonian mechanics and special relativity mechanics from the genetic viewpoint, i.e. from the point of view of how relativity theory "grew out of" Newtonian mechanics, electrodynamics and classical optics (Frank, 1946, 1971: p. 19). This approach, which by the way is within the bounds of "diachronic

logic",[5] may have been one of the reasons why Frank saw the two theories as logically and empirically comparable rather than "incommensurable". For on this approach Einstein generalised and combined certain components of classical physics previously thought to be independent of each other, viz. the principle of relativity and the principle of constancy of light velocity (independence of source velocity); the replacement of Galilean transformation by Lorentz transformation to relate space and time measurements in inertial frames accounts for the conceptual changes mentioned before, since the former was based on two assumptions rejected in special relativity mechanics, viz. "(1) the time interval (time) between two events is independent of the condition of motion of the body of reference, (2) the space interval (distance) between two points of a rigid body is independent of the condition of motion of the body of reference" (Einstein, 1916, 1964: p. 30) and since the Einsteinian modification of the concept of mass was sufficient to make the laws of mechanics covariant with respect to Lorentz transformation between inertial frames. Frank's view of the logical relation between Newtonian mechanics and special relativity mechanics was thus in line with the tradition, started by Einstein himself, according to whom the special theory of relativity "...has... been developed from electrodynamics as an astoundingly simple combination and generalisation of the hypotheses, formerly independent of each other, on which electrodynamics was built" (Einstein, 1916, 1964: p. 41). According to the same tradition, although both special and general theories of relativity "... possessed a decidedly revolutionary appearance when they were announced, it has now become clear that they represent the natural termination for the classical theories of mechanics and electromagnetism, rather than a break with these systems of ideas and the inception of a new line of thought" (Lawden, 1971: p. viii).

5. When discussing conceptual changes in the transition from Newtonian mechanics to special relativity mechanics Frank did not use any clear-cut conception of meaning. He did approve of the basic ideas of operationalism, in particular of the requirement that a (metric) concept in physics, to be empirically meaningful, has to be associated with measurement operations which (within limits of experimental error) give unambiguous results. However he did not

follow Bridgman in claiming that distinct methods of measurement, e.g. of distance, time interval, temperature, etc., yield different concepts, of space, time, temperature, etc. (Bridgman, 1927: pp. 66–91), which could have formed—under suitable additional assumptions—a premiss for concluding that Newtonian mechanics and special relativity mechanics were conceptually completely disparate or "incommensurable". Nor did he appeal for that purpose to the syntactic difference between the concepts of Newtonian mechanics and special relativity mechanics, quite rightly, so it seems, since one can always relativise Newtonian concepts in an inessential way by introducing strictly redundant terms.

Questions of definability (e.g. of Newtonian "mass" in terms of relativistic "mass") and of translation (e.g. from Newtonian mechanics to special relativity mechanics) cannot be reliably answered in an intuitive, informal context, i.e. without more rigorous formulation of the theories in question and without conventions concerning the analytic components (meaning-postulates) of those theories. It is possible, therefore, that—under suitable assumptions—the language of special relativity mechanics may be shown to be closed, in Ajdukiewicz's sense (1934), with respect to the theoretical sub-language of Newtonian mechanics, i.e. the former cannot be enriched with the concepts of the latter without either modifying some of the concepts or making the enriched language disconnected.

Notes

* First published in *Br. J. Phil. Sci.* **24**, 270–6, Aberdeen University Press, 1973.
1. I have traced the history of the problem in the writings of conventionalist philosophers in Giedymin (1978).
2. Cf. Frank, (1946, 1971): Chap. 2.
3. Cf. similar claims in Einstein 1916, 1964: pp. 24, 27, 30; Grünbaum, 1954; Feynman et al., 1966: Chap. 16, p. 2.
4. Einstein, (1916, 1964): Chaps. 5 and 13.
5. Cf. Suszko, 1968.

Bibliography

ABRAHAM, R. and MARSDEN, J. E. (1967) *Foundations of Mechanics*, Benjamin, New York-Amsterdam.
AJDUKIEWICZ, K. (1921, 1966) Z metodologii nauk dedukcyjnych, Lwow [From the Methodology of Deductive Sciences, (translated by J. GIEDYMIN)], *Studia Logica*, **XIX**, 9–46.
AJDUKIEWICZ, K. (1934) Das Weltbild und die Begriffsapparatur, *Erkenntnis*, **54**, 259–87 (English translation in (1978)).
AJDUKIEWICZ, K. (1936) Empiryczny fundament poznania [The Empirical Foundation of Knowledge], *Spraw. Poznań. Tow. Przyj. Nauk*, 27–31.
AJDUKIEWICZ, K. (1948, 1976) Metodologia i metanauka, *Życie Nauki*, Methodology and Metascience, in *25 Years of Logical Methodology in Poland* (Edited by M. PRZEŁECKI and R. WÓJCICKI), Reidel, Dordrecht.
AJDUKIEWICZ, K. (1949, 1973) Zagadnienia i kierunki filozofii, Czytelnik, Warszawa, *Problems and Theories of Philosophy* (translated by H. SKOLIMOWSKI and A. QUINTON), Cambridge University Press.
AJDUKIEWICZ, K. (1953) W sprawie artykułu prof. A. Schaffa o moich poglądach filozoficznych [A reply to Prof. A. Schaff's article concerning my philosophical views], *Myśl Filoz.*, **2**, 8, 292–334; reprinted in *Language and Knowledge*, Vol. II, 176.
AJDUKIEWICZ, K. (1960, 1965) *Język i Poznanie [Language and Knowledge, Selected Papers]*: 1920–1939, Vol. I, PWN, Warsaw, 1945–1963, Vol. II, PWN, Warsaw.
AJDUKIEWICZ, K. (1965, 1974) *Pragmatic Logic* (translated by O. WOJTASIEWICZ), Reidel and PWN, Dordrecht and Boston.
AJDUKIEWICZ, K. (1978) *Kazimierz Ajdukiewicz: The Scientific World-Perspective and Other Essays 1931–1963* (edited by J. GIEDYMIN, Reidel, Dordrecht.
  (1931) On the meaning of expressions, pp. 1–34.
  (1934) Language and meaning, pp. 35–66.
  (1934) The world-picture and the conceptual apparatus pp. 67–89.
  (1935) On the problem of universals, pp. 95–110.
  (1935) The scientific world-prespective, pp. 111–17.
  (1936) Syntactic connexion, pp. 118–39.
  (1937) A semantical version of the problem of transcendental idealism, pp. 140–54.
  (1947) Logic and experience, pp. 165–81.
  (1948) Epistemology and semiotics, pp. 182–91.
  (1960) Axiomatic systems from the methodological point of view, pp. 282–94.
  (1964) The problem of empiricism and the concept of meaning, pp. 306–19.
  (1967) Intensional expressions, pp. 320–47.
  (1967) Proposition as the connotation of a sentence, pp. 348–61.

BERTHELOT, R. (1911) Un romantism utilitaire, *Étude sur le mouvement pragmatiste*, Félix Alcan, Paris.
BETH, E. W. (1959) *The Foundations of Mathematics*, North-Holland, Amsterdam.
BLANCHÉ, R. (1955, 1962) *Axiomatics*, Routledge & Kegan Paul, London.
BOHNERT, H. (1965) Craig's theorem, *Journal of Philosophy*, **62**, 10, 251-81.
BOLTZMANN, L. (1897, 1974) *Lectures on the Principles of Mechanics*; reprinted as Part 4 of *Theoretical Physics and Philosophical Problems* (edited by B. MCGUINNESS), Reidel, Dordrecht.
BONOLA, R. (1955) *Non-Euclidean Geometry*, Dover Publications, New York.
BOOLE, G. (1854, 1951) *An Investigation of the Laws of Thought*, Dover Publications, New York.
BORN, M. (1956) Physics and relativity, in *Physics In My Generation*, Pergamon, London and New York.
BORN, M. (1962) *Einstein's Theory of Relativity*, Dover Publications, New York.
BORN, M. (1971) *The Born-Einstein Letters*, Macmillan, London.
BRAITHWAITE, R. (1953, 1959) *Scientific Explanation*, Cambridge University Press.
BRIDGMAN, P. (1927) *The Logic of Modern Physics*, Macmillan, New York.
CAMPBELL, N. R. (1921) Theory and experiment in relativity, *Nature*, **106**, 804-6.
CARNAP, R. (1936/7) Testability and meaning, *Philosophy of Science*, **3**, 419-71, and **4**, 1-40.
CARNAP, R. (1937) Foundations of logic and mathematics, *International Encyclopedia of Unified Science*, **3**, Chicago University Press.
CARNAP, R. (1952) Meaning postulates, *Philosophical Studies*, **3**, 65-73.
CARNAP, R. (1956) *Meaning and Necessity*, Appendix, Meaning postulates, Chicago University Press.
CARNAP, R. (1958) Beobachtungssprache und theoretische Sprache, *Dialectica*, **12**, 236-48.
CARNAP, R. (1966) *Philosophical Foundations of Physics*, Basic Books, New York.
CAYLEY, A. (1845) On certain results relating to quaternions, *Philosophical Magazine*, **CLXXI**, 141-45.
CAYLEY, A. (1858) Report on the recent progress of theoretical dynamics, *Rep. Br. Ass. Advmt. Sci.*, 27th meeting.
CAYLEY, A. (1859) A sixth memoir upon Quantics, *Phil. Trans. R. Soc.*
CAYLEY, A. (1884) Presidential address, *Rep. Br. Ass. Advmt. Sci.* 53rd meeting.
CLARK, R.W . (1973) *Einstein, the Life and Times*, Hodder and Stoughton, London.
COOLIDGE, J. L. (1963) *A History of Geometrical Methods*, 1-451, Dover Publications, New York.
CORNMAN, J. (1972) Craig's theorem, Ramsey-sentences and scientific instrumentalism, *Synthèse*, **25**, 82-127.
CRAIG, W. (1953) On axiomatisability within a system, *J. Symbolic Logic*, **18**, 30-2.
CRAIG, W. (1956) Replacement of auxiliary expressions, *Phil. Review*, **65**, 38-55.
CUNNINGHAM, E. (1910) The principle of relativity in electrodynamics and an extension thereof, *Proc. Lond. Math. Soc.*
CUVAJ, C. (1968) Henri Poincaré's mathematical contributions to relativity and the Poincaré stresses, *American Journal of Physics*, **36**, 12, 1102-13.
D'ABRO, A. (1939, 1951) *The Rise of the New Physics*, Dover Publications, New York.
DARBOUX, G. (1873) *Sur une classe remarquable de courbes et des surfaces algébriques*, Gauthier-Villars, Paris.
DARBOUX, G. (1914) Éloge historique d'Henri Poincaré, *Mém. Acad. Sci. Inst. Fr.*, **52**.

DESCARTES, R. (1901) *Discourse on Method* (translated by J. VEITCH), L. Walter Dunne, London.
DESCARTES, R. (1954) *The Geometry of René Descartes* (translated by D. E. SMITH and M. L. LATHEM). Dover Publications, New York.
DUHEM, P. (1894) Quelques réflexions au sujet de la physique expérimentale, *Revue Quest. Scient.*, 2nd series, **III**.
DUHEM, P. (1906, 1974) *La théorie physique: son objet, sa structure* (translated by P. P. WIENER), *The Aim and Structure of Physical Theory*, Atheneum, New York.
DUHEM, P. (1908, 1974) The Value of Physical Theory, *Rev. Gén. Sci. pur. appl.*, **XIX**, reprinted as Appendix in *The Aim and Structure of Physical Theory*, Atheneum, New York.
DUHEM, P. (1908, 1969) ΣΩzein ta Φainomena: *Essai sur la notion de théorie physique de Platon à Galilée*, Paris, A. Hermann et Fils, Paris, *To Save the Phenomena*, translated by E. DOLAND and C. MASCHLER. Chicago University Press.
EHRENFEST, P. T. (1959) *Collected Scientific Papers*, North-Holland, Amsterdam.
EINSTEIN, A. (1905, 1967) Über einen die Erzeugung und Verwandlung des Lichtes betreffenden heuristischen Gesichtspunkt, *Annln. Phys.*, Bd. **17**, Fol. 4, 132–48, On a heuristic point of view about the creation and conversion of light, in D. TER HAAR, *The Old Quantum Theory*, Pergamon, Oxford.
EINSTEIN, A. (1905, 1970) On the electrodynamics of moving bodies, *Special Theory of Relativity*, pp. 187–218 (edited by C. W. KILMISTER), Pergamon, Oxford.
EINSTEIN, A. (1916, 1964) *Relativity: The Special and General Theory*, Methuen, London.
EINSTEIN, A. (1954) *Ideas and Opinions*, Crown Publishers, New York.
EULER, L. (1748) Réflexions sur l'espace et le temps, *Histoire de l'Académie des Sciences et Belles Lettres*, Berlin.
FEYERABEND, P. K. (1964) Realism and instrumentalism, *The Critical Approach to Science and Philosophy*, pp. 280–308 (edited by M. BUNGE), Collier-Macmillan, New York.
FEYNMAN, R., LEIGHTON, R. and SANDS, M. (1963) *The Feynman Lectures on Physics*, Reading (Mass.), Addison-Wesley.
FORD, L. R. (1929, 1951) *Automorphic Functions*, 2nd. ed., Chelsea Publishing Co., New York.
FOURIER, J-B. J. (1822, 1878) *The Analytical Theory of Heat* (translated by A. FREEMAN), CUP, London.
FRANK, P. (1946, 1971) *Foundations of Physics*, International Encyclopedia of Unified Science. Chicago University Press.
FRANK, P. (1948) *Einstein: His Life and Times*, Jonathan Cape, London.
FRIEDMAN, M. (1953) The methodology of positive economics, *Essays in Positive Economics*, Chicago University Press.
GEACH, P. T. (1970) A program for syntax, *Synthèse*, **22**, 1/2.
GERGONNE, J. D. (1818) Essai sur la théorie des définitions, *Annls. Math.*, **IX**.
GIEDYMIN, J. (1968) Revolutionary changes, non-translatability and crucial experiments, in *Problems of the Philosophy of Science*, 223–7 (edited by A. MUSGRAVE and I. LAKATOS), Vol. III of *Int. Coll. in Philosophy of Science*, North-Holland, Amsterdam.
GIEDYMIN, J. (1970) The paradox of meaning variance, *Br. J. Phil. Sci.*, **21**, 257–68.
GIEDYMIN, J. (1976) Instrumentalism and its critique: a reappraisal, *Boston Studies in the Philosophy of Science*, **XXXIX**, 179–207.
GIEDYMIN, J. (1977) On the origin and significance of Poincaré's conventionalism, *Studies in History and Philosophy of Science*, **8**, 4.

GIEDYMIN, J. (1978) Radical conventionalism, its background and evolution: Poincaré, LeRoy and Ajdukiewicz, Introductory essay in *Kazimierz Ajdukiewicz: the Scientific World-Perspective and Other Essays, 1931-1963*, Reidel, Dordrecht.
GLYMOUR, C. (1971) Theoretical realism and theoretical equivalence, *Boston Studies in the Philosophy of Science*, **VIII**, 275-88.
GOLDBERG, S. (1967) Henri Poincaré and Einstein's theory of relativity, *Am. J. Phys.*, **35**, 934-44.
GOODMAN, N. (1955, 1965) *Fact, Fiction and Forecast*, 2nd ed., The Bobbs-Merrill Co., Indianapolis and New York.
GRASSMAN, H. (1844) *Ausdehnugslehre*, O. Wigand, Leipzig.
GRAVES, R. P. (1885) *Life of Sir W. R. Hamilton*, 3 vols., Dublin.
GRÜNBAUM, A. (1954) Operationalism and relativity, in *The Validation of Scientific Theories*, pp. 83-92 (edited by P. FRANK), Collier Books, New York.
GRÜNBAUM, A. (1963) R. Carnap's views on the foundations of Geometry, in *The Philosophy of Rudolf Carnap* (edited by P. SCHILPP), Open Court, La Salle, Illinois.
GRÜBAUM, A. (1963, 1973) *Philosophical Problems of Space and Time*, 2nd ed., Reidel, Dordrecht.
GRÜNBAUM, A. (1968) *Geometry and Chronometry in Philosophical Perspective*, University of Minnesota.
HADAMARD, J. (1914/1921) L'oeuvre Mathématique de Poincaré, *Acta Math.* **38**, 203-87.
HAMILTON, W. R. (1931) *The mathematical Papers of Sir William Rowan Hamilton*, Vol. I, Geometrical Optics, (edited by A. W. CONWAY and J. L. SYNGE), Cambridge. C.U.P.
(1828) Theory of systems of rays, pp. 1-106.
(1830) First supplement to an essay on the theory of systems of rays, pp. 107-44.
(1837) Third supplement to an essay on the theory of systems of rays, pp. 164-293.
(1831) On a view of mathematical optics, pp. 295-6.
(1833a) On some results of the view of a characteristic function in optics, pp. 297-303.
(1833b) On a general method of expressing the paths of light and of the planets, by the coefficients of a characteristic function, pp. 311-32.
Editors' Appendix, Note 14, on the relation of Hamilton's optical method to dynamics, pp. 484-7.
Editors' Appendix, Note 19, on the transition from the emission theory to the wave theory, pp. 497-9.
Editors' Appendix, Note 20, on group velocity and wave mechanics, pp. 500-2.
HAMILTON, W. R. (1940) Vol. II, *Dynamics* (edited by A. W. CONWAY and A. J. MCCONNELL).
(1834a) On a general method in dynamics; by which the study of the motions of all free systems of attracting or repelling points is reduced to the search and differentiation of one central relation, or Characteristic Function, pp. 103-60.
(1834b) On the application to dynamics of a general mathematical method previously applied to optics, pp. 212-16.
(1834) Second essay on a general method in dynamics, pp. 162-212.

HAMILTON, W. R. (1967) Vol. III, *Algebra* (edited by H. HALBERSTAM and R. E. INGRAM).
— (1833c) Theory of conjugate functions, or algebraic couples; with a preliminary and elementary essay on algebra as the science of pure time, pp. 3–100.
— (1847) On the application of the method of quaternions to some dynamical questions, pp. 441–54.
— (1854) Preface to "Lectures on quaternions", pp. 117–55.
HANSON, N. R. (1958) *Patterns of Discovery*, Cambridge University Press.
HEERDEN, B. J. VAN (1968) On the history of the theory of relativity, *Am. J. Phys.*, **36**, 1171–2.
HELMHOLTZ, H. (1868, 1968) Über die Tatsachen die der Geometrie zum Grunde liegen, *Nachr. Königl. Ges. Wiss. zu Göttingen*. Also in *Populärwissenschaftliche Vorträge*.
HELMHOLTZ, H. (1870, 1968) On the origin and significance of geometric axioms, (translated by E. ATKINSON) in *Philosophy of Science: The Historical Background* (edited by J. KOCKELMANS), The Free Press, New York.
HEMPEL, C. (1945) Geometry and empirical science, *Am. Math. Mon.*, **52**.
HEMPEL, C. (1958) The theoretician's dilemma, *Minn. Stud. Phil. Sci.*, **II**.
HEMPEL, C. (1963) Implications of Carnap's work for the philosophy of science, in *The Philosophy of R. Carnap*, 685–707 (edited by P. SCHILPP) Open Court, La Salle, Illinois.
HERTZ, H. (1890) Über die Grundgleichungen der Electrodynamik für ruhende Körper, *Annln. Phys. Chem.*, Neue Folge, **XL**, 8, 577–624.
HERTZ, H. (1894, 1956) *The Principles of Mechanics* (translated by D. E. JONES and J. T. WALLEY), Macmillan, London.
HESSE, M. B. (1963) *Models and Analogies in Science*, Sheed and Ward, London and New York.
HESSE, M. B. (1976) Duhem, Quine and a new empiricism, in *Can Theories Be Refuted? Essays on the Duhem–Quine Thesis* (edited by S. G. HARDING), Reidel, Boston and Dordrecht.
HESSE, M. B. (1980) *Revolutions and Reconstructions in the Philosophy of Science*, Harvester Press, Brighton.
HILL, E. L. (1951) Hamilton's principle and the conservation theorems of mathematical physics, *Rev. Mod. Phys.*, **23**, 3.
HOFFMANN, B. (1975) *Einstein*, Paladin, Frogmore, St. Albans.
HOLTON, G. (1960, 1974) On the origins of the special theory of relativity, in *Thematic Origins of Scientific Thought: Kepler to Einstein*, 2nd printing pp. 165–84, Harvard University Press, Cambridge, Massachusetts.
HOLTON, G. (1964, 1974) Poincaré and relativity, *Thematic Origins*, ibid., 185–95.
KAHAN, T. (1959) Sur les origines de la théorie de la relativité restreinte, *Revue Hist. Sci. Applic.*, **12**.
KESWANI, F. H. (1965–6) Origin and concept of relativity, Part I, *Br. J. Phil. Sci.*, **15**, 286–306, Part II, **16**, 19–32, Part III, **16**, 273–94.
KLEIN, F. (1872) Vergleichende Betrachtungen über neuere geometrische Forselungen; A. Düchert, Erlangen.
KLEIN, F. (1910) Über die geometrischen Grundlagen der Lorentzgruppe, *Jber. Dt. Mat. Verein.*, **XIX**, 1 Abt. Heft. 9/10, 281–300.
KLEIN, F. (1926, 1957) *Vorlesungen über höhere Geometrie*, 3 Aufl. Bearbeitet und

herausgegeben von W. Blashke, Reprinted Chelsea Publishing Company, New York.
KLEIN, F. (1926/7, 1956) *Vorlesungen über die Entwicklung der Mathematik im 19. Jahrhundert, (1926)* Bd. I. Bearbeitet von R. COURANT and O. NEUGEBAUER (1927) Bd. II. Bearbeitet von R. COURANT and ST COHN-VOSSEN, reprinted Chelsea Publishing Company, New York.
KLEIN, F. (1928, 1959) *Vorlesungen über Nicht-Euklidische Geometrie*, bearbeitet von W. ROSEMANN, Berlin, reprinted Chelsea Publishing Company, New York.
KOENIGS, G. (1897) *Leçons de Cinématique* Professées à la Sorbonne. Cinématique théorique, Paris, A. Hermann, Paris.
LAKATOS, I. and MUSGRAVE, A. (eds.) (1970) *Criticism and the Growth of Knowledge*, Cambridge University Press.
LARMOR, J. (1894) A dynamical theory of the electric and luminiferous medium, *Phil. Trans. R. Soc.* Series A, **185**, Part II, 719–822.
LARMOR, J. (1900) *Aether and Matter*, Cambridge. C.U.P.
LAUE, M. VON (1911) *Das Relativitätsprinzip*, T. Vieweg, Braunschweig.
LAWDEN, D. (1971) *Tensor Calculus and Relativity*, Methuen, London.
LEROY, E. (1899) Science et philosophie, *Revue de Métaphysique et de Morale*, **VII**.
LEROY, E. (1900–1) La science positive et les philosophies de la liberté, *Rapports presentés au Congrès Internationale de Philosophie*, Paris.
LEROY, E. (1901) Un positivisme nouveau, *Revue de Métaphysique et de Morale*, **IX**.
LEWIS, D. (1970) General semantics, *Synthèse*, **22**, 1/2.
LIE, S. (1871, 1959) On a class of geometric transformations, in *A Source Book in Mathematics* (edited by D. E. SMITH), Dover Publications, New York.
LIE, S. and ENGEL, F. (1888–1893) *Theory der Transformationsgruppen*, B. G. Teubner, Leipzig.
LOBATCHEVSKY, N. (1840, 1955) *Untersuchungen zur Theorie der Parallel-Linien*, English translation in R. BONOLA, *Non-Euclidean Geometry*, Dover Publications, New York.
LORENTZ, H. A. (1892) La théorie électromagnetique de Maxwell et son application aux corps mouvants, *Archs. Néerl. Sci.*, **25**, 363; reprinted in *Collected Papers*, 1935–9, Vol. 2, pp. 164–343, Nijhoff, The Hague.
LORENTZ, H. A. (1895, 1935–39) *Versuch einer Theorie des elektrischen und optischen Erscheinungen in bewegten Körpern*, Leiden; reprinted in *Collected Papers*, Vol. 5, pp. 1–137, Nijhoff, The Hague.
LORENTZ, H. A. (1899) Théorie simplifiée des phénomènes électriques et optiques dans des corps en mouvement, *Versl. K. Akad. Wet. Amst.*, reprinted in *Collected Papers*, Vol. 5, pp. 139–55, Nijhoff, The Hague.
LORENTZ, H. A. (1904, 1970) Electromagnetic phenomena in a system moving with any velocity less than that of light, *Proc. Sect. Sci. K. Ned. Akad. Wet.*, Amsterdam, **VI**, 1903/4, Part II, July 1904, 809–30; reprinted in *Special Theory of Relativity* pp. 119–44 (edited by C. W. KILMISTER), Pergamon, Oxford.
LORENTZ, H. A. (1909, 1915, 1952) *The Theory of Electrons and Its Applications to the Phenomena of Light and Radiant Heat*, Dover Publications, New York.
LORENTZ, H. A. (1915/1921, 1954) Deux mémoires de Henri Poincaré sur la Physique Mathématique, *Acta Math.*, **38**, 293; *Collected Papers*, Vol. 7, pp. 258–73; *Oeuvres de Henri Poincaré*, **IX**, pp. 683–701, Notes et Commentaires, Gauthier-Villars, Paris.
LORENTZ, H. A. (1920) *The Einstein Theory of Relativity: A Concise Statement*, Brentano, New York.

LUKASIEWICZ, J. (1970) *Selected Works*, North-Holland, Amsterdam, London and Warszawa.
MCCORMMACH, R. (1967) Henri Poincaré and quantum theory, *Isis*, **58**, 37–55.
MCCORMMACH, R. (1970a) H. A. Lorentz and the electromagnetic view of nature, *Isis*, **61**, 459–97.
MCCORMMACH, R. (ed.) (1970b) Einstein, Lorentz and the electron theory, *Historical Studies in The Physical Sciences*, Second Annual Volume, Pennsylvania University Press.
MCCREA, W. H. (1957) Edmund Taylor Whittaker, *J. Lond. Math. Soc.* **32**, 234-56.
MCCREA, W. H. (1978) Ebenezer Cunningham, *Bull. Lond. Math. Soc.*, **10**, 116–26.
MCKINSEY, J. C. C., SUGAR, J. and SUPPES, P. (1953) Axiomatic foundations of classical particle mechanics, *J. Rat. Mech. Analysis*, **2**, 253–72.
MCKINSEY, J. C. C. and SUPPES, P. (1955) On the notion of invariance in classical mechanics, *Br. J. Phil. Sci.*, **5** (Feb.), 290–302.
MARTIN, D. (1976) Whittaker, E. T., *Dictionary of Scientific Biography*, Vol. **XIV** Scribners, New York.
MILHAUD, G. (1896) La science rationelle, *Revue de Métaphysique et de Morale* (May).
MILLER, A. I. (1975) A study of Henri Poincaré's "Sur la dynamique de l'Electron", *Archs. Hist. Exact Sci.*, **10**, 207–328.
MINKOWSKI, H. (1907) Die Grundgleichungen für die elektromagnetischen Vorgänge in bewegten Körpern, *Nachr. Königl. Ges. Wiss. zu Göttingen.* 53–111.
MINKOWSKI, H. (1907, 1915) Das Relativitätsprinzip, *Annln. Phys.*, **47**, 927–38.
MINKOWSKI, H. (1908, 1922) Raum und Zeit, in *Das Relativitätsprinzip*, (edited by A. SOMMERFELD), Teubner, Leipzig.
MORRIS, C. (1938) Foundations of the theory of signs, *International Encyclopedia of Unified Science*, **1**, 2, Chicago University Press.
NADEL-TURONSKI, T. (1976) Zasady fizyki jako metaprawa, *Studia Metodologiczne*, Poznan.
NAGEL, E. (1961) *The Structure of Science*, Routledge, London.
NOETHER, E. (1918) Invariante Variationsprobleme, *Nachr. Königl. Ges. Wiss.*, Math. Phys. Klasse, Heft 2, Göttingen.
PEIRCE, C. (1958–60) *Collected Papers*, Harvard University Press, Cambridge, Massachusetts.
PLÜCKER, J. (1835) *System der analytischen Geometrie*, Duncker and Humblot, Berlin.
PLÜCKER, J. (1846) *System der Geometrie des Raumes in neuer analytischen Behandlungsweise*, W. H. Scheller,
PLÜCKER, J. (1865) On a new geometry of space, *Phil. Trans. R. Soc.*, 725–91.
PLUMMER, H. C. (1910) On the theory of aberration and the principle of relativity, *Mon. Not. R. Astr. Soc.*, LXX, 3, 252–75.
POINCARÉ, H. (1881) Sur les fonctions fuchsiennes, *C. R. Hebd. Séanc. Acad. Sci.*, **92**, **93**.
POINCARÉ, H. (1887) Sur les hypothèses fondamentales de la géométrie, *Bull. Soc. Math. Fr.*, **XV**, 203–16.
POINCARÉ, H. (1889) *Leçons sur la théorie mathématique de la lumière*, Naud, Paris.
POINCARÉ, H. (1890, 1901) *Electricité et optique: les théories de Maxwell et la théorie électromagnetique de la lumière*, (edited by J. BLONDIN), G. Carré, Paris.
POINCARÉ, H. (1891, 1902, 1952) Les géométries non-euclidiennes, *Revue Gén. Sci. Pur. Appl.*, **2**, 769–74, Reprinted in *Science and Hypothesis*, pp. 35–50.
POINCARÉ, H. (1892–99, 1951) *Méthodes nouvelles de la mecanique celeste*, Gauthier-Villars, Paris; Dover Publications, New York.

POINCARÉ, H. (1895, 1954) A propos de la theorie de M. Larmor, Parts I–IV, Éclair. Élect., Tomes 3 and 5, *Oeuvres*, **IX**, pp. 369–426, Gauthier-Villars, Paris.
POINCARÉ, H. (1896–7) Sur les solutions périodiques, et le principe de moindre action, *C. R. Hebd. Séanc. Acad. Sci.*, **123**, 224–8; **124**, 227–30, *Oeuvres*, **IX**.
POINCARÉ, H. (1897) Les idées de Hertz sur la méchanique, *Rev. Gén. Sci. Pur. Appl.*, **8**, 734–43, *Oeuvres*, **IX**, pp. 231–50 (partly reprinted in *Science and Hypothesis*, pp. 231–50, 1902, 1952.
POINCARÉ, H. (1898, 1905) De la mesure du temps, *Revue de Métaphysique et de Morale*, **VI**, 1–13, reprinted in *The Value of Science*, 1905, 1958.
POINCARÉ, H. (1898a) On the foundations of geometry, *Monist*, **IX**, 1, 1–43.
POINCARÉ, H. (1899) Des fondements de la géométrie, A propos d'un livre de M. Russell, *Revue de Métaphysique et de Morale*, **VII**, 251–79.
POINCARÉ, H. (1900, 1902, 1952) Sur les rapports de la physique expérimentale et de la physique mathématique, *Rapports presentés au Congrès Internationale de Physique*, Paris, 1900, Tome I, pp. 1–20, reprinted in *Science and Hypothesis, pp.* 140–82, 1902, 1952.
POINCARÉ, H. (1900a) La théorie de Lorentz et le principe de réaction, *Archs. Néerl. Sci.*, **5**, 252–78, *Oeuvres* **IX**, pp. 464–88.
POINCARÉ, H. (1901) Sur une forme nouvelle des équations de la méchanique, *C. R. Hebd. Séanc. Acad. Sci.*, **132**, 369–71.
POINCARÉ, H. (1901, 1902, 1952) Sur les principes de la méchanique, *Bibl. du Congrès Internationale de Philosophie*, Paris, 1900–1, reprinted in *Science and Hypothesis*, 1902, 1952.
POINCARÉ, H. (1901) *Electricité et Optique*, Paris.
POINCARÉ, H. (1902, 1952) *La Science et l'Hypothèse*, E. Flammarion, Paris, *Science and Hypothesis*, Dover Publications, New York.
POINCARÉ, H. (1902) Sur la valeur objective des théories physiques, *Revue de Métaphysique et de Morale*, reprinted in *The Value of Science*, 1905, 1958.
POINCARÉ, H. (1902a, 1956) Les fondements de la géométrie, *Bull. Sci. Math.* Sept. 1902, *Oeuvres* **XI**, pp. 92–113.
POINCARÉ, H. (1904, 1905, 1958) L'état actuel et l'avenir de la physique, International Congress of Arts and Sciences at St. Louis, *Bull. Sci. Math.*, **28**, 302–24; reprinted in *The Value of Science*, Chaps. VII–IX, 1905, 1908; English translation in The progress of mathematical physics, *Monist*, **15**, 1, 1–24, 1905.
POINCARÉ, H. (1905a, 1954) Sur la dynamique de l'electron, *C. R. Hebd. Séanc. Acad. Sci.*, **140**, 1504–8, 5 juin 1905; *Oeuvres* **IX**, pp. 489–94.
POINCARÉ, H. (1905, 1958) *La Valeur de la Science*, E. Flammarion, Paris; *The Value of Science*, Dover Publications, New York.
POINCARÉ, H. (1906, 1954, 1970) Sur la dynamique de l'electron, *Rc. Circ. Mat. Palermo*, **21**, 129–176; *Oeuvres* **IX**, pp. 494–550; *Special Theory of Relativity*, Introduction and 1–4, 9, pp. 145–86, (edited by C. W. KILMISTER), Pergamon, Oxford.
POINCARÉ, H. (1908, 1956) *Science et Méthode*, E. Flammarion, Paris; *Science and Method*, Dover Publications, New York.
POINCARÉ, H. (1909) Sur une généralisation de la méthode de Jacobi, *C. R. Hebd. Séanc. Acad. Sci.*, **149**, 1105–8.
POINCARÉ, H. (1908a, 1954) La dynamique de l'electron, *Revue Gén. Sci. Pur. Appl.* **19**, 386–402; *Oeuvres* **IX**, pp. 551–86.
POINCARÉ, H. (1911, 1954) Sur la théorie de quanta, *Oeuvres* **IX**, pp. 620–5.
POINCARÉ, H. (1912, 1954) Sur la théorie de quanta, *Oeuvres* **IX**, pp. 626–53.

POINCARÉ, H. (1912, 1954, 1963) L'hypothèse des quanta, *Oeuvres* **IX**, pp. 654–68; (English translation) Quantum theory, *Last Essays* pp. 75–88, 1963.
POINCARÉ, H. (1913, 1920, 1963) *Dernières Pensées*, E. Flammarion, Paris; *The Last Essays*, Dover Publications, New York.
POINCARÉ, H. (1954) *Oeuvres de Henri Poincaré*, Tome **IX**, Gauthier-Villars, Paris.
POLANYI, M. (1958) *Personal Knowledge*, Routledge, London.
PONCELET, J. V. (1822) *Traité de proprietés projectives des figures*, Paris.
POPPER, K. R. (1935, 1959) *The Logic of Scientific Discovery*, Hutchinson, London.
POPPER, K. R. (1956, 1963) Three views concerning human knowledge, *Contemporary British Philosophy*, 3rd series (edited by H. D. LEWIS); reprinted in *Conjectures and Refutations*, Routledge, London.
POPPER, K. R. (1963) *Conjectures and Refutations*, Routledge, London.
POPPER, K. R. (1966) A note on the difference between the Lorentz–Fitzgerald contraction and the Einstein contraction, *Bri. J. Phil. Sci.*, **16**, 64, 332–3.
POPPER, K. R. (1972) *Objective Knowledge: an Evolutionary Approach*, Oxford University Press.
PRZEŁĘCKI, M. (1969) *The Logic of Empirical Theories*, Routledge, London.
PRZEŁĘCKI, M. (1974) A set-theoretic versus a model-theoretic approach to the logical structure of physical theories, *Studia Logica*, **33**, 91–112.
PRZEŁĘCKI, M. and WØJCICKI, R. (eds) (1976) 25 *Years of Logical Methodology in Poland*, Reidel, Dordrecht.
PUTNAM, H. (1965, 1979) Craig's theorem, *J. Phil.* **62**, 10; reprinted in *Mathematics, Matter and Method*, pp. 228–37, Cambridge University Press.
PUTNAM, H. (1975, 1979) The refutation of conventionalism, *Mind, Language and Reality*, pp. 153–91, Cambridge University Press.
QUINE, W. (1936, 1949) Truth by convention, *Readings in Philosophical Analysis* (edited by H. FEIGL and W. SELLARS, Appleton-Century-Crofts, New York.
QUINE, W. (1953) Two dogmas of empiricism, *Phil. Rev.*, **60**, 2043.
RAMSEY, F. P. (1978) *Foundations: Essays in Philosophy, Logic, Mathematics and Economics* (edited by D. H. MELLOR), Routledge, London.
REICHENBACH, H. (1944, 1965) *Philosophic Foundations of Quantum Mechanics*, California University.
REICHENBACH, H. (1921, 1959) The present state of the discussion on relativity, *Modern Philosophy of Science* (translated by M. Reichenbach), pp. 1–45, Routledge, London.
RIEMANN, B. (1854, 1959) Über die Hypothesen welche der Geometrie zugrunde liegen, *Abhandlungen der Gesellschaft der Wissenschaften zu Göttingen*, Bd. 13; reprinted in H. WEYL, *Das Kontinuum*, Leipzig, Von Veit, 1918; On the hypotheses which lie at the foundations of geometry, in *A Source Book in Mathematics* (edited by D. E. SMITH), Dover Publications, New York.
ROSEN, E. (ed.) (1939, 1959) *Three Copernican Treatises*, (translated with Introduction, Notes and Bibliography), 2nd. ed., Dover Publications, New York.
RUSSELL, B. (1897, 1956) *An Essay on the Foundations of Geometry*, Dover Publications, New York.
SCHAFFNER, K. (1969) The Lorentz theory of relativity, *Am. J. Phys.* **37**, 498–513.
SCHEFFLER, I. (1960) Theoretical terms and a modest empiricism, in *Philosophy of Science*, pp. 159–76 (edited by A. DANTO and S. MORGENBESSER), The World Publishing Company, Cleveland.

SCHEFFLER, I. (1969) *The Anatomy of Inquiry*, Routledge, London.
SCHEIBE, E. (1979) On the structure of physical theories, *The Logic and Epistemology of Scientific Change* (edited by I. NIINILUOTO and R. TUOMELA, *Acta Philosophica Fennica*, **XXX**, 2–4, North-Holland, Amsterdam.
SCHLICK, M. (1979) *Philosophical Papers*, Vol. I, 1909–72 (edited by H. MULDER and B. VAN DER VELDE), Reidel, Dordrecht.
   (1915) The philosophical significance of the principle of relativity, pp. 153–89.
   (1917) Space and time in contemporary physics. An introduction to the theory of relativity and gravitation, pp. 207–69.
SCRIBNER, C. (1964) Henri Poincaré and the principle of relativity, *Am. J. Phys.* **32**, 672–78.
SEELIG, C. (1954) *Albert Einstein*, Europa Verlag, Zürich.
SHAPERE, D. (1965) *Philosophical Problems of Natural Science*, Macmillan, New York.
SITTER, W. DE (1911) On the bearing of the principle of relativity on gravitational astronomy, *Mon. Not. R. Astr. Soc.*, **LXXI**, 388–415.
SNEED, J. D. (1971) *The Logical Structure of Mathematical Physics*, Reidel, Dordrecht.
STEGMÜLLER, W. (1976) *The Structure and Dynamics of Theories*, Springer, New York.
SUPPES, P. (1957) *Introduction to Logic*, Van Nostrand, New York.
SUPPES, P. (1969) A comparison of the meaning and uses of models in mathematics and the empirical sciences, in P. SUPPES, *Studies in the Methodology and Foundations of Science*, pp. 10–23, Reidel, Dordrecht.
SUSZKO, R. (1968) Formal logic and the development of knowledge, in *Problems of the Philosophy of Science* pp. 210–22 (edited by A. MUSGRAVE and I. LAKATOS), Vol. III of *Int. Coll. in Philosophy of Science*, North-Holland, Amsterdam.
SYNGE, J. L. (1937) *Geometrical Optics: Introduction to Hamilton's Method*, Cambridge University Press.
TARSKI, A. (1933, 1936) *Pojecie prawdy w jezykach hauk dedukcyjnych*, Towarzystwo Naukowe Warszawskie, Warsaw; (German translation) Der Wahrheitsbegriff in formalisierten Sprachen, *Studia Philosophica*, I.
TEMPLE, G. (1956) Edmund Taylor Whittaker, *Biogr. Mem. Fellows R. Soc.*, **2**, 299–323.
TOULMIN, S. (1961) *Foresight and Understanding*, Indiana University Press, Bloomington.
TUOMELA, R. (1973) *Theoretical Concepts*, Library of Exact Philosophy, Springer-Verlag, Wien, New York.
TUOMELA, R. (1978) On the structuralist approach to the dynamics of theories, *Synthèse*, **39**, 211–31.
VOIGT, W. (1887) Über das Doppler'sche Prinzip, *Nachr. Königl. Ges. Wiss. zu Göttingen*.
WANG, H. (1957) The axiomatization of arithmetic, *J. Symbolic Logic*, **22**, 152.
WHITTAKER, E. T. (1904, 1937) *A Treatise On The Analytical Dynamics of Particles and Rigid Bodies*, C.U.P., Cambridge.
WHITTAKER, E. T. (1910) Recent researches on space, time and force, *Mon. Not. R. Astr. Soc.*, **LXX**, 363–6.
WHITTAKER, E. T. (1940) The Hamiltonian Revival, *Mathl. Gaz.* **XXIV**, 260, 153–8.
WHITTAKER, E. T. (1945) The sequence of ideas in the discovery of quaternions, *Proc. R. Ir. Acad.*, **50**, 93–8.
WHITTAKER, E. T. (1953, 1973) *A History of the Theories of Aether and Electricity*, vol. 2.

WHITTAKER, E. T. (1955) Albert Einstein, *Biogr. Mem. Fellows R. Soc.*, **1**, 37–67.
WILBOIS, J. (1899) La méthode des sciences physiques, *Revue de Métaphysique et de Morale* (May).
WILLIAMS, C. P. (ed.) (1968) *Relativity Theory: its Origins and Impact on Modern Thought*, R. E. Krieger Publishing Company, New York.
WILLIAMS, P. M. (1973) On the logical relations between expressions of different languages, *Br. J. Phil. Sci.*, **24**, 357–67.
ZAHAR, E. (1973) Why did Einstein's programme supersede Lorentz's, *Br. J. Phil. Sci.*, **24**, Part I, 2, 95–123; Part II, **24**, 3, 223–62.

Name Index

Abraham, M.   153, 170
Abraham, R.   50
Adrastus   51
Ajdukiewicz, K.   xiii–xv, 33–6, 38–9, 41, 52, 109–45, 192, 197–8, 205
Alexander, P.   147
Argand, J. R.   66
Aristotle   51, 92, 97, 110

Bacon, F.   101
Bellarmino   92, 94–6, 98–100, 103, 137
Beltrami, E.   71
Bergson, H.   3, 118, 124, 127, 147
Berkeley, G.   45, 92, 94, 110
Bernard, C.   119
Berthelot, R.   78
Beth, E.   148
Bohnert, H.   43
Bohr, N.   92, 196
Boltzmann, L.   82
Bolzano, B.   111
Bonola, R.   38, 146
Boole, G.   87
Born, M.   39, 150–3, 158–61, 165, 191
Bosanquet, B.   41, 147
Bouquet, J.-C.   180
Boutroux, E.   3, 119
Bradley, F. H.   41, 147
Brahe, T.   101
Braithwaite, R.   42–3, 196
Bridgman, P.   42, 92, 196, 199, 205
Bruns, H.   66

Campbell, N.   162–4
Cantor, G.   113
Carnap, R.   10, 15, 17, 32, 38, 48, 116, 137, 196, 198, 202–3

Cauchy, A. L.   xii, 78
Cayley, A.   vii, 18, 36, 64, 67, 73, 76–7, 84, 166, 180
Chasles, M.   39
Clark, R. W.   150
Clausius, R.   39
Cohen, M. R.   97
Comte, A.   119
Conway, A. W.   59, 85
Coolidge, J. L.   77
Copernicus, N.   95, 97–8, 101
Coulomb, Ch. A.   200
Couturat, L.   39
Craig, W.   43
Cunningham, E.   191, 193
Curie, M.   190

D'Abro, A.   88
Darboux, G.   vii, 36, 84, 89, 178, 180
Darwin, C.   12
Dedekind, J. W. R.   39, 146
Dercyllides   51
Descartes, R.   22, 53, 55, 86
Dilthey, W.   111
Drabkin, I. E.   97
Duhem, P.   vii, x, xii, 3–4, 9, 16, 26, 37–8, 40, 44, 50–2, 62, 80, 92, 94–5, 98, 100–3, 105–8, 115, 119, 126–7, 144, 147, 192

Eddington, A. S.   92
Einstein, A.   viii, xi–xiii, 37, 39, 40, 88, 149–54, 157–64, 166, 169–71, 186, 189–93, 196, 200, 203–5
Euclid   ix, 23, 41, 61, 109, 113–14, 117, 126, 131, 143–4, 146, 178–9
Euler, L.   174, 194

217

Name Index

Fermat, P. 28, 53–4, 58
Feyerabend, P. K. 43, 92, 94–6, 98–9, 101, 196
Feynman, R. 150, 205
Fleck, L. viii
Ford, L. R. 70, 88, 176
Foscarini, P. A. 98
Fourier, J.-B. J. vi, 44, 78, 80, 174
Frank, P. viii, 150, 191, 198–205
Fresnel, A. 27, 54, 60, 81, 156, 173, 181–3, 186–8

Galileo, G. x, 51–2, 100–3, 114, 204
Galois, E. vii
Gauss, C. F. 20, 38, 67
Geach, P. T. 147
Geminus 51, 97
Gergonne, J. D. 23, 25
Gibbs, J. W. 12–13
Giedymin, J. 64–5, 146–8, 197, 205
Gödel, K. 146
Goldberg, S. 150, 152, 170–1, 186
Grassmann, H. 67
Grünbaum, A. 6–12, 15–17, 20, 40, 146 205

Hadamard, J. viii
Halberstam, H. 86
Hamilton, W. R. xii, 27–8, 44–6, 53–68, 70–84, 174–184
Hanson, N. R. 196
Harding, S. G. 38, 40
Heerden, P. J. van 150
Helmholtz, H. von 25, 38, 185
Hempel, C. 43, 196
Hermite, C. vii, 82, 92
Herschel, J. 86
Hertz, H. 12–13, 63, 82, 92, 179–80, 184–5
Hesse, M. B. 16, 38, 40, 50, 193
Hilbert, D. xiv, 68
Hipparchus 51
Hoffmann, B. 150, 190
Holton, G. viii, 5, 37, 43, 150–2, 167, 169–72, 192–4
Hume, D. 37, 110
Husserl, E. 111
Huyghens, C. 54

Ingram, R. 86

Jacobi, C. 74, 78, 80

Kahan, T. 150, 152
Kant, I. viii, xiv, 2, 10, 37, 61–2, 110–11, 113–16, 128, 139, 177, 191
Kaufmann, W. 35, 37, 116, 169, 172
Kepler, J. 52, 103
Keswani, G. H. 150, 193
Kilmister, C. W. 72, 88, 150
Kirchhoff, G. 63, 92
Klein, F. vii, x, 22, 29, 39, 65, 72, 75–6, 78, 146, 166, 180, 191
Koenigs, G. xii, 179–80
Kotarbinski, T. xiv, 110
Kuhn, T. S. 40, 196

Lagrange, J. L. vi, 44, 53–4, 62, 73, 78, 80, 174
Lakatos, I. 43
Lange, F. xiv, 2
Laplace, P. S. 78, 174
Larmor, J. xii, 46–7, 159, 182–3, 185, 190–1, 193
Lawden, D. 204
Leibniz, G. 53
LeRoy, E. xiii, xiv, 3–4, 18, 33, 37–8, 40, 111–13, 116, 118–29, 139, 147, 192, 197
Lewis, D. 147
Lie, S. vii, ix, x, 21–3, 25, 29, 38–9, 41, 46, 65, 75–6, 78, 114, 180
Lloyd, H. 54, 64
Lobatchevsky, N. 16, 20, 23, 25, 39, 178
Lorentz, A. xi, 29, 35, 37, 72, 114, 149–66, 168–70, 172–6, 180, 185–6, 188–94, 204
Łukasiewicz, J. xiii, xiv, 52

McCormmach, R. 150
Mach, E. xi, 42, 45, 92, 94, 110, 191
McCrea, W. 46, 191, 193
Malus, E.-L. 59
Marsden, J. E. 50
Martin, D. 151

Maupertuis, P. L. M. de 28, 54, 58, 72
Maxwell, J. C.   vi, 27, 31, 47, 63, 69–70, 72, 78–82, 114, 147, 154, 156 170–1, 181, 183–4, 186–8, 203
Mellor, D. H. 85
Merton, R. K. xiii
Michelson, A. A. xi, 153, 156, 194
Milhaud, G. 3, 119, 122, 147
Mill, J. S. 37, 136
Miller, A. J. xii, 150–2, 169–72, 186
Minkowski, H. xiii, 72, 146, 150, 152–3, 157–8, 165–6, 191
Monge, G. 75
Morley, F. W. xi
Morris, C. 140

Nadel-Turonski, T. 89
Nagel, E. 16, 43, 81, 196
Neumann, F. 27, 39, 181–3, 186
Newton, I. ix, x, 17, 34, 39, 53–4, 68, 72, 94, 124, 131, 138, 179, 191, 196–205
Nicod, J. 48
Nietzsche, F. 2
Noether, E. 88

O'Connor, D. J. 147
Ohm, M. 62
Osiander, A. 92, 94, 96–100, 103, 137
Ostwald, W. 80

Peacock, G. 62, 66
Peano, G. 39, 146
Peirce, C. 52, 107
Plato 37, 51
Plücker, J. vii, 12, 22, 25, 29, 39, 64 76–7, 178
Plummer, H. C. 165, 191
Poincaré, H. vii–xiv, 1–21, 23–9, 32–42, 44–7, 63–5, 68–84, 90, 92, 94, 108–9, 111–19, 123–8, 131–2, 137, 139, 144, 146–7, 149–62, 165–6, 168–70, 197
Poisson, S. D. xii, 78, 80
Polanyi, M. 43
Poncelet, J. V. vii, 23, 77, 166, 180
Popper, K. R. viii, 5, 16, 43, 52, 81, 92–8, 116, 150, 196

Proclus 52
Przełecki, M. 43, 134, 199
Ptolemy, C. 51, 101
Putnam, H. 43

Quine, W. 32–3, 38, 40–1, 107, 116, 125–6, 144, 148, 191
Quinton, A. 148

Ramsey, F. P. x, xii, 42–50, 53–60, 62–3, 163, 198, 202
Rankine, W. J. M. 80
Reichenbach, H. viii, 10, 15–17, 150, 167, 191–2, 196
Riemann, B. 6–10, 16–21, 25, 38, 39
Rodrigues, O. 67
Rosen, E. 97, 99, 101
Russell, B. 17–18, 39, 41, 48, 147

Schaff, A. 133–4, 146
Schaffner, K. 150–2, 162, 165, 170–1
Scheffler, I. 43
Schilpp, P. 170
Schlick, M. viii, 163–4, 191, 193
Schrödinger, E. 85
Scribner, C. 150
Seelig, C. 154
Shapere, D. 196
Simmel, G. 148
Simplicius 51
Sitter W. de 165, 191
Skolimowski, H. 148
Smith, D. E. 39
Sneed, J. D. 43, 50
Spranger, E. 111, 148
Stegmüller, W. 43
Steiner, J. 39
Suppes, P. 43
Suszko, R. 205
Synge, J. L. 59, 72

Tannery, P. 180
Tarski, A. 140–1
Temple, G. 46
Theon 51

Name Index

Toulmin, S. 43, 196
Tuomela, R. 43, 147–8

Urban VIII 103

Vaihinger, H. xiv
Voigt, W. 191

Wang Hao 146
Warren, J. 66
Watson, G. N. 42

Weber, W. 39
Weyl, H. 16, 38
White, M. 116
Whittaker, E. xi, xii, 45–6, 65, 68, 75, 77, 85, 149–54, 161–2, 165, 167, 169, 173, 191, 193–4
Wilbois, J. 3, 119, 147
Williams, C. Pearce 150
Williams, P. M. 148
Wittgenstein, L. 193
Wojcicki, R. 134

Zahar, E. 150

Subject Index

Absolute motion 177–8, 185, 194; impossibility to detect, as a general law (Poincaré) 185; space 177–9, elimination of, by differentiation (Poincaré) 179; time 177–8

Acceptance, Assertion meaning-rules (Ajdukiewicz) 33–6, 129–31, 134–6; *vs.* rejection 131

Action and the meaning of science (LeRoy) 4, 123–4; and foresight 124; philosophy of 4, 123, 127; primacy of (LeRoy) 123

Aether, *see* **Ether**

Amorphousness metric, of space 7–9, 20, 25, 38

Analogy and mathematical discovery 70–2; and the discovery of quaternions 66; as the basis of applications of analysis in physics 69; multiplication in arithmetic and in the calculus of quaternions 69; between wave optics and Hamilton's dynamics 72; formal 44, 69–70, 72

Analysis and physics 69; epistemological status of, and algebra (Hamilton) 61–3; the Kantian view of the foundations of, 3

Analytic geometry 53, 55; judgement 19, 32, 35, 116; statement as convention 35; analytic–synthetic distinction 32–3, 41, 147, questioned by Quine 116 and Russell 147; Ajdukiewicz language without analytic statements 144; logical tautologies 35; analyticity 116; relativised to language 116, 138–9, 142

A Priori form of sensibility and of understanding 114; synthetic 11, 19, 24, 30–1, 62, 140; axiom of math. induction as, 24, 113; apriorism 32, 192, Kantian 11, 32, 37, radical *vs.* moderate 111, *vs* empiricism 138, 145

Assertion, *see* **Acceptance** assertive *vs.* non-assertive axiomatic systems in methodological sense (Ajdukiewicz) 135, 147; hypotheses as unassertible (instrumentalism in the pragmatic sense) 102, 175

Axiom(s) and an Ajdukiewicz language 34, 134; and the Ramsey view of theories 47, 49; as implicit definitions 3, 9, 15, 23–4, 39; of arithmetic 24, 39; of free mobility 11, 19; of geometry as conventions 2, 24; of infinity and of choice as hypotheses 135; of a language 33–4; of math. induction 24, 146

axiomatic system in the pragmatic *vs.* formal sense (Ajdukiewicz) 134–7, assertive-deductive 135, assertive-reductive 135, neutral 135, 137, 147, uninterpreted *vs.* interpreted 134, 136; axiomatic meaning rule 24, 34, 35, 137, 147–8; axiomatic system 34, 134; meaning-postulate 15, 24

Choice conventional 12; and simplicity 12, 14, 19, 106; simplification of problems 25; guided by empirical considerations 32, 35, 115; of congruence class 8, 26; of coordinate system 2, 19, 22, 25, 87; of geometry 2, 16, 19, 25, 114; of hypothesis 19, 106, 115; of theory of language and meaning 144–5; of language 36; of laws, based on pragmatic criteria 106; of space elements 12, 22, 64; of space dimensionality 64, 76, 114

221

Subject Index

Classification natural (Duhem) 9, 43, 147; no natural classn. exists (LeRoy) 121–2
Commodism 112, 147
Conceptual apparatus (Ajdukiewicz) 125, 133, and radical conventionalism 143, and "truth" 133, 139, and world-picture 133; disparity between Lorentz's and Einstein's electrodynamics 163–4, between Newtonian and relativity mechanics 191, 193, 196–205; framework 118; and ontological interpretation of a formalism 163
Content of a theory iv; as invariant under change of conventions (Poincaré) 30–1, 83, 174; cognitive 31, 75, 82–3, 108, 151, 160; common to rival theories 27, 59, 160, 175; empirical 27, 30, 59, 163, 175; of a physical law, varying with applications (Duhem) 106; of a theory and the Ramsey view 49–50, 163; as observational consequences and the form of equations (Poincaré) 63, 75, 82, 108; sameness of, 108, 163
Conventionalism
  Duhem's c. Duhem theses 16–17, 26, 40, 101, 104–5, 144; multiplicity of observationally indistinguishable or equivalent explanation 79, 102; natural classification 9, 43, 147; instrumentalism 50, 90, 94–5, 98, 102–5, in astronomy 52, 96–7, 100–6, restricted 53, 60, 97; ambiguity of observations 40; theory as experimental laws and uninterpreted calculus 47–50; paradox of experimental method 100–2
  extreme c. (nominalism) of LeRoy x, and pan-theoreticism and the problem of the existence of the universal invariant 3, 5, 16, 38, 118–28; laws as expressing conventions of measurement 121; Poincaré's criticism of, 123–4, 126–7; correspondence and knowledge 122, as substitutivity of objects 122; the critique of facts 119–20, of laws 120–1 of theories 122–3; the new critique of science 3–4, 30, 119, 122–3; action and the meaning of science 4, 123; primacy of action 123, philosophy of action 4, 123, 127; importance of rival theories 122; no true theory in the limit 122
  moderate c. of Poincaré 44, 75, 77, 79, 83, 180, 183, 192 and the principle of relativity 29, 159–61, 170 and the Plücker–Lie principle of transformation 22–3; and evolutionary epistemology; and the pr. of equivalence 160; and the discovery of relativity theory, vii, xi–xiii, 186–90; and Koenigs's theorem xii, 179–80, 183; and Lorentz's research programme 168–9, 173–4; and the principle of theoretical tolerance 189–90, 195; conventional and empirical components of a theory separated 116, conventional status of "absolute time", "absolute space", "simultaneity of distant events", and of Euclidean geometry in classical mechanics 114, 178–9; conventionality of dimensionality 12–13, 64, 76; elimination of "absolute space" by differentiation 179; empirical generalisation elevated to the status of conventions 25, 27; geometric conventionalism 2, 6–10, 13, 20–3, 25–30, 113, 176; generalized conventionalism 3, 6, 9–10, 27–30, 47; group-theoretic approach to geometry 19, 21–3, 114, group admitted by an equation x, 29, 83, 75–6; generalization of apparently rival theories 84; epistemological problem, main ix, 84, 174–5, 187; indifferent hypotheses as interchangeable conventions 30, 187, invariantism ix, and limit of knowledge 14–5, 174–5; philosophy of the physics of the principles xii, 84, 90, 189–90, 195; structuralism and structuralist realism xi, 14–15, 174–5, 182, 189; Poincaré-metric 8, Poincaré–LeRoy dispute 4; Riemann-Poincaré thesis of the conventionality of congruence 7–9
  radical c. of Ajdukiewicz xiii–xiv, 36, 40, 109–45, and methodological pluralism 145; and the existence of not intertranslatable languages 40, 109, 127, 142–3; choice of object language *vs.* choice of a concept of language 143–5; dependence of knowledge of the choice of language 110; disruptive nature of scientific changes 110; compared with logical positivism 110, and the (French) new critique of science 110–11, and Kantian philosophy 111, and the philosophy of Dilthey 111, and understanding

Subject Index 223

methodology of mathematics xiv; incomparability of world-pictures associated with different conceptual apparatuses 133–4, 137–8; humanistic understanding and the evolutionary viewpoint 133–4; distinction between observational and interpretative statements rejected 127; problems not solvable without convention 118; thesis of radical c., weak and strong 132–3
Coordinate definition, *see* **Definition**
Correspondence, *see* **Duality, Transformation**

Definition explicit (equivalence) 23, 43, 48; implicit (by postulates), 3, 9, 15, 23–4, 39, 113, 116, analogous to equations 23; and elimination of theoretical terms 43; and proofs of existence and uniqueness 24; coordinative 7–8, 15, 34; in disguise 2, 23
Dictionary and the Ramsey view of theories 47–9; and translation 8–9, 23, 34, 84, 126
Discovery of automorphic functions vii, 70–2, 176; of Hamilton's method in optics 54–9, 72; of Maxwell's theory 184; of quaternions 65–9, 72; of relativity theory 149–205, debate over the d. of, 149–52, 169–73, Born's views 152–4, 158, Einstein's view 154, 161, Holton's view 167–8, Lorentz's view 154–9, 161, Whittaker's view 149–51, 161–2, 167, 173; simultaneous d. of, by Poincaré and Einstein 161, 188–90
Displacement(s) 14, 25, 114, group of 25, 114
Duality (Reciprocity) 23, 27, 39; of geometric (projective) axioms 23, 65, 77, of geometric terms 23; Plücker's clarification of 77, principle of, in projective geometry 39; algebraic sense of 39; interchange of categories of space elements and contact transformations 65; between Fermat principle of least time and Maupertuis's principle of least action in optics 28, 58 and dynamics 59; parallelism 28

Elimination 42–5, 48, 80; of theoretical terms 42–3; by (explicit) definition 43, 60; replacement method (Craig's) 88, (Ramsey's) 42, 48, 60, 88; the eliminant (Ramsey) 48; by mathematical principles 80; Hamilton's method in geometrical optics as replacement method 44, 59–60; the physics of the principles as replacement programme 80; eliminability of theoretical terms 60; of metaphysics 42, 80
Empiricism and elimination programmes 44, 80; genetic (Poincaré) 113, 116; geometrical, *see* **Geometry**; logical, 32, 43, 140, 167, 192–3; and the distinction between context of discovery and of justification 167; moderate 6, 9–10, 16, 18, 32, 37, 144, new 38, 40; orthodox (traditional) 6, 30, 106; Poincaré's criticism of geometric 3, 11, 13, 17–18, 20–1, 25–6; Duhem's criticism of traditional 106, extreme (radical) 37, 42, 80, 111, 144; operationalism 42
  empiricist philosophy 1, 3, 43, 113–14, of physics 28, 31, 113
  empirical criteria and choice of metric geometry 26; and language change 35; content 17, 27, 59; *vs.* conventional element of knowledge 33; generalisations 25, 28, 32, elevated to status of conventions (Poincaré) 25, 32, 44, 83, 116, 144; meaning-rule of an Ajdukiewicz language 34
Epistemology apriorist 44; conventionalist, *see* **Conventionalism**; empiricist, *see* **Empiricism**; evolutionary 12, 113; of the physics of the principles 45, 78–82; elimination of, in the socio-historical view of science 144–5; of Ajdukiewicz, Duhem, LeRoy and Poincaré, *see* **Conventionalism**; hermeneutics 111; limits of knowledge 14, 29, 45, 81; objectivity 81; pluralist 111; semantical 140; Hamilton's 60; Kantian 44; realist and the semantical concept of truth 141
  epistemological aspect of dispute over absolute motion 45; role of group theory

78; controversies and the theory of language 138, 140, 143–5; interdependence of geometry and physics 17; nature of the concept of content 151; Poincaré's main problem 5, 83–4, 174; sense of instrumentalism 45; change in status of statements 32, 138; status of relativity principle 116, 169–71; status of observationally indistinguishable laws 107; elaboration of the Riemann–Poincaré pr. of conventionality of congruence 7

Equivalence ambiguity of, in non-formal sense 164; as sameness of historical tradition 167; between Lorentz's "new mechanics" and special relativity, the problem of 151, 155, 162–3, 193; as "continuous with" 162, 168; between Lorentz's and Einstein's electrodynamics (Schlick) 163–4; as experimentally indistinguishable 103, 107, 122; functional 41; physical e. principle of, and the pr. of relativity (Poincaré) 160; observational e. 13–14, 51–2, 102, 107, 146, 161, 163–4, observational *vs.* conceptual 161–8, multiplicity of obs. equivalent hypotheses for any set of obs. data 102, 122; conceptual non-equivalence (disparity) 163–4; non-e. between Poincaré's and Einstein's (special) principle of relativity claimed (Holton) 170–1.

Ether (Aether) invented to give laws of mechanics the form of differential equations (Poincaré) 173; and electrons as the only "true matter" (Lorentz) 169; and Poincaré's physics and philosophy 172, 176, 186; effects of, on optical experiments and abs. velocity 185, and field equations 188; in Fresnel's optics 181–2, 187; in Lorentz's 1892, 1895 and 1904 electrodynamics 157, 168, 185; in Neumann's optics 181–2; hypothesis of the e. and the principle of relativity (de Sitter) 165

Evolution theory of 2, 41; evolutionary epistemology 12; and radical conventionalism 133–4

Experiment crucial 15, 26, 93–4, 101, 103, 108, impossibility of 16, 101, 103; and geometry 11, 15, 20; Kaufmann's1906 e. 35, 37, 169 and the pr. of special relativity 35, 116

experimental ambiguity 16; method, Duhem's view of, contrasted with Galileo's 101; limits of exptl method (Duhem) 102–3, 108; paradox of exptl method (Duhem) 102, 105; results (facts) systematised by a mathematical principle 27, 44; experimentally test(able) 15, 26, 32; exptly equivalent and indistinguishable hypotheses 103, 147

Facts common-sense 4, 120, 123–4; experimental 3, 19, 44; pure ("bare", "rough") 120, and science 124–5; created by scientists (LeRoy) 4, 120, 124–5 (Ajdukiewicz) 12; *vs.* theory 4, 30, 82, 125; and conventions 124; and positive science (Comte) 119; as "intersections of laws" (LeRoy) 121; as dependent on theories 30, 50; relations between crude facts invariant under change of scientific conventions (Poincaré) 126; and interpretations (Poincaré, Duhem, Ajdukiewicz—polemic) 125

Falsification avoidance of, 26–7, 40, 94, 121; and Duhem thesis 26, and Poincaré thesis 26; falsificationism 5, 93

Form of a theory 63, 82–3; and math. discovery 70; common to several theories 79, 174–5; of an equation x, 83, 156; and invariants 158; and group admitted by equation x, 29, 75, 78, 183; of sensibility 86, 114; of understanding 86, 114; Ramsey-sentence f. 81; *vs.* material of a group (Poincaré) 77;

formal similarity 70, 72, 175; and discovery of automorphic functions 70

formalism of wave optics extended to dynamics (Hamilton) 72, the four-vector formalism in Poincaré (1905/6) 158

Geometry and experiment 11, 16, 20; and Koenigs's kinematics 180; as a study of invariants under transf. groups viii, 29, 65, 76, 114; choice of 2, 22; Erlanger Programm

Subject Index 225

146, 166; interdependence of g. and physics 17, 40; Kant's philosophy of 3, 20, 113; Lie's sphere g. 22, 65; Lobatchevskian 19–20, 39, 69 and Darboux transfs 178; metric g. as intertranslatable 27, 34, 113, 177; physical *vs.* pure 9–11, 15, 17; Plücker's line g. 22, 65, Plücker's philosophy of 77; Poncelet's programmes 77, 166; projective approach to g. and the relationship between classical and relativistic mechanics (Klein) 86, 166; geometrisation of physics 29, 65; quasi-geometric approach to physics 46, 65, 78, 82, 95–6
Group vii–xi, 11–12, 19–21, 25, 29, 36, 71, 114, group-theoretic approach to geometry 19, 21, 114; group–theory as a basis of mathematics and physics 77–8; Klein's Erlanger Program 78, 146, 166; group admitted by a differential equation (Lie) 29, 75 and the form of equations 83; Galileo–Newton group 24, 114, 178; the Galileo–Newton and Lorentz groups as subgroups of the affine g. (Klein), 84, 166; the Lorentz group 29, 72, 114, 157, 166, and its invariants (Poincaré 1905) 158–9, the inhomogeneous Lorentz g. (the Poincaré g.) 190; non-reciprocal Lorentz transformations do not form a g. 168, g. of transformations 21, 29, 114, and displacements 25, 114, g. of continuous transformations and Koenigs's kinematics 180, g. of linear fractional transformation and solutions of linear homogeneous differential equations 72, 75–6, g. of canonical (contact) transformation 75, physics as a study of invariants under transformation groups 29, 114
See also **Invariants, Relativity, Transformation**

Hermeneutic(s) 5, 111, 167–8, 193; and meaning 167–8, 193; philosophy of Dilthey 111; and empathy (humanistic understanding) 193; and historical analysis 168, and unmasking 168, 193
Holistic view of theories 40, 50, 107
Hypothesis ad hoc, in Lorentz's theory (Poincaré) 156–7; auxiliary, and Duhem's thesis 26, 50, impossibility of testing isolated h. 16; indifferent h. (Poincaré) 187; as mathl. instruments 102–3; astronomical h. 51–2, 97–8, 103, and the Platonic principle 57, and the method of astronomy 51–2, 96–7, implicit in the use of differential equations in physics (Boltzmann) 82;
hypothetico–deductive system as assertive–reductive (Ajdukiewicz) 135
hypotheticism (fallibilism) 93, 95
h-free interpretation of Lorentz transformations (Poincaré) 100–1

Incommensurability between classical and the "new mechanics" overcome within the projective approach (Klein) 166; between Lorentz's and Einstein's electrodynamics (Schlick) 163–4, as resulting from contradictory ontological assumptions (Schlick) 164; thesis xiii, 95, 192, not inter-translatable theories (Ajdukiewicz) 192; conceptual disparity 163, 193
Instrumentalism Duhem's 50, 90–108, and Duhem's theses 98, 102–5, and replacement programmes 43–5, 80–1; *vs.* essentialism and hypotheticism 92, *vs.* physics of the principles 81, *vs.* realism ix, 52, 95–6, 99–100, 106, Maxwell's 147; ontological, semantical, pragmatic and epistemological senses of, 81, 91, 96, 102–6, 137, 148, 182 and the Ramsey-sentence 43–4, 47–50; restricted 53, 60, 97; in astronomy 51–3; i. sense of Lorentz's 1904 transformations 157; and neutral axiomatic systems (Ajdukiewicz) 83, 135, 137
Invariance of deductive relations under translation 34; of total scope of meaning-rules under exchange of expressions (Ajdukiewicz) 130–1, 141–3; Lorentz-i. of the laws of physics 154 and the special pr. of rel. 35, 92, 114, 156; analytical dynamics and i. under canonical transformations 74, 76

Subject Index

invariants   integral 30; invariance principles and the Ramsey-sentence 81; physics as a study of invariants 65, 114; theory of the i. of the affine group 166, classical, relativistic mechanics and the affine group (Klein) 166; rel. theory as theory of i. relative to a transformation group (Klein) 166; under theoretical change 4, 37, 118; "universal invariant" (Poincaré–LeRoy controversy) 118, 125–6

Isotopes   in the language-matrix (Ajdukiewicz) 130, 136

Knowledge   as correspondence (LeRoy) 122; and foresight 124; and language (Ajdukiewicz) 109, 128, 138; and models (Duhem) 147; as relational (Poincaré) x, 32, 187; diverse orders of (LeRoy) 123; in psychological and logical sense (Ajdukiewicz) 129; limit of (Poincaré) 14, 29, 45, 91, 103–8; and evolutionary epistemology 12

Language(s)   Ajdukiewicz language 134, not intertranslatable 34, 109, 125–6, 128, 136, 138–9, 144, 148, closed *vs.* open 130–2, 136, 138, 142–4, matrix 130–1, 137, 139, 143; connected and disconnected 130, 132–3, 142; linguistic world-perspectives 128; universal 133; and radical conventionalism 128–38, 142–3; change of 26–7, 35; intertranslatable and the problem of a universal invariant (Poincaré) 125–6, 176

Laws   elevated to conventional principles (Poincaré) 116, invariant under change of theoretical language (Poincaré) 115, 175–6; experimentally indistinguishable 106; of geometrical optics and the wave and particle theories of light 59; form-invariance of physical laws 29, 114, 176; statistical laws characteristic of modern physics (Poincaré) 174; scientific *vs.* commonsense (Duhem) 103; as conventions 4–5, underdetermined by evidence (Duhem) 104–5, as instruments 93, as symbolic 104; presupposed in measurements (Poincaré) 124; the ether and laws as differential equations (Poincaré) 173; and the Ramsey-sentence 48–9; and Duhem thesis 107

Meaning   Ajdukiewicz's theory of 24, 33–5, 109, 128–39, 142, 148; conceptions of m. and cognitive role of language 128, 138; and logical semantics 141; denotation 136, 140–2, co-denotation 142; theory dependence of 192; neo-Kantian conception 115; as the class of observational consequences 167; pragmatist concept of 115; and the Ramsey view 49; determination and hermeneutics 167

Measurement   of space 7–8, 15, of time 3, 114, measuring rod as a rigid body 7–8; intrinsic metric 7, 17; laws presupposed 104–6, 125; limited sensitivity of measuring instruments 104; re-metrisation 7–8, 16–18

Metaphor   as model 5; change of metaphors 115; mathematical m. 107

Model   and metaphor 5, 31–2, 115; as realisation 59, 63; mathematical 53, mechanical 46, 50, 79, Duhem's view of 50

Natural   classification (Duhem) 9

New mechanics   of Lorentz 151–3, 159, 169; of Lorentz, Einstein, Poincaré and Minkowski (Klein) 166, and projective approach to geometry (Klein) 166

Nominalism   of LeRoy 3–4, 16, 27, 83, 118–28, anti-empiricist 17–18; *vs.* realism 83; Poincaré's criticism of LeRoy's 123–4, 126–7; nominalist attitude 26–7, 40, 83, 127

Objectivity   5, 83–4, 120, 123–4, 174, and facts (LeRoy) 120, 126; main epistemological problem (Poincaré) 174; of science defended (Poincaré) 126–7; of science questioned (LeRoy, Duhem) in comparison with religion 126; and the "universal invariant" 121, 123, 126

Observational (numerical) equivalence 13, 41, 51–2, 59, 76, 83, 102, 107–8, 115–16, 146; metric geometries as o. equivalent (Poincaré) 27; o. indistinguishable theories and laws 29, 83, 104–5, 107, 115; o. and theoretical terms distinction 43–4, 93; o. equivalence of the Lagrangian and Hamiltonian dynamics 76; o. consequences and Ramsey-sentence 88; multiplicity of o. equivalent hypotheses (Duhem and Poincaré) 102, 104–5, 115; common o. part of rival theories 59, of the particle and wave theories of light 59

observation, ambiguity of 40, theory dependent 40

Pan-theoreticism 3–4, theory-dependence of facts (LeRoy) 4, laws dependent on theories (Duhem) 102–8

Phenomena saving the, 5, 50–1; and hidden mechanism 79–81; *vs.* noumena 62

phenomenalism 42–3, 62, 80, 82, 110, of logical positivists 110, of Mach and Russell 42

phenomenalist philosophy 80; phenomenological theory in physics 80, 88

Positivism 3, 42–3, 80; logical 110, of Berlin and Vienna Circles 110, and radical conventionalism 110; new French (LeRoy) *vs.* Comtian 123; contemporary anti-positivism 167, and debate over origins of relativity 167

Pragmatic conception of language 109, 112, 148; view of axiomatic systems 112, 134, 147; sense of instrumentalism 100–6; *vs* syntactical and semantical 83

pragmatism of LeRoy 120, 123; and conventionalism 76–8, 81–2, 107, 127

Principle(s) *a priori*, as constitutive of objects of knowledge 63; conventional, and change of language 35; of equivalence 83; of duality 39, 77, algebraic sense of 39, 77; empirical generalisations elevated to conventional principles 25, 27, 32, 44, 116, 131, 144; extremal (variational) 46, 53, 56–7, 59, 73–4, 88; Fermat's 28, 54, 58–9, in optics and dynamics 59, 79; general prs. and common content of rival theories 44, 66; Hamilton's 3, 27, 79, in dynamics 27, 73, in optics 27; Maupertuis's of least action 28, 54, 57–9, in optics and dynamics 59, 79; of relativity (Einstein's) 151, 170–1, 186, 191, (Galileo–Newton's) 194, (Poincaré's) 115, 157, 172, 175–6, 178, 185–6, 188–9, 192; physics of the principles 174, *vs.* instrumentalism 81, neutral with respect to rival hypotheses 174, and its philosophy 42–84, 174

Priority debate over discovery of special relativity 151, 153, 161–2

Quaternions, *see* **Discovery**

Ramsey-sentence of a theory 42, 49, 53, 60, 81, 88, 194 and inter-theory relations 49–50, the content of a theory, 49–50, 81; second order language 49; primary and secondary system 47–9, 52, 60; uninterpreted calculus 50, and observational consequences 88; variational principle in optics (Hamilton's) as the Ramsey-sentence 53, 59–60

Realism naïve 91, 96, 99–100, 169, structural (Poincaré) 83

realist interpretation of Fitzgerald contraction (Lorentz) 168–9; r. view of theories 52, 82–3, 102

Reciprocity, *see* **Duality**

Reduction group–theoretic in physics 84, 166

228   Subject Index

Relativity of geometry 177, of motion 115, 178–9, of space 12, 39, 177, 185
  principle of   Einstein's special 151, 191, Galileo–Newton 194, Poincaré's 115, 157, 175, 178, 185–6, 189, 192, first formulation (Poincaré) in 1895 185, mathematics of relativity in Poincaré's 1905–6 188–9; constancy of light velocity in inertial frames 186, 188–9; covariance of Maxwell's equations under Lorentz transformations 72
  relativity theory of the affine group as generalisation of classical and relativistic mechanics (Klein) 166
  See also **Discovery, Groups, Invariance, Principle(s), Transformations**

Save   the phenomena 5, 28, 50, 52, 97–8, 100–3, 108 and Duhem's argument 101; the Platonic principle 51, 97–8
Scepticism   and instrumentalism 137, and new critique of science 123
Simplicity   12, 14, 19, 20, 25, 64, 114; of Einstein's special rel. *vs.* Lorentz's (Schlick) 163; of a group 19; of hypotheses 14, 20, 99, 105, and choice of metric geometry 25; problems and transformations 29, 76–7; and dimensionality of space 64
Space   amorphousness of 8–9, 20, 38, 113; as mathematical continuum 177; Hamilton's view on 85; s. element 12, 22, 64, 178; extended 12, *vs.* local s. 12; homogeneity of 10; infinity of 11, relativity of 12, 39, 177; nature of, and Lorentz transformations (Einstein) 154; structural properties of 9–10; symmetry of 66, 85; metrisation of, alternative 8–9, 20, 38, 113, 177; Kant's philosophy of 10–11; relational view of 7, 9, 12
  absolute s.   and time 154, 178, to be abandoned in new mechanics (Poincaré) 188, 194, as conventions 194, reality of (Euler) 194
  hyper-space   12, 62, 64–5, 72, 76; representative, visual tactile, motor 10
  space-time   72
Structure(s)   of theory 31, 42, 44, 48, 63, 78, 88; of language 115, 129, 137 and meaning 129; language matrix (Ajdukiewicz) 130, 137; groups as 77–8; expressed by equations 82, and the content of a theory 82, 108; common of rival theories 44 and elimination programmes 44; similarity of 63, 108, 175 and unification of theories 63.
Symmetry   and invariance 70, 72, generalised coordinates and momenta 73, of physical laws 72, of space 66, 85
Synthetic *a priori*   11, 41, 62

Theorem(s)   and the Ramsey view 47; Koenigs's in kinematics 179–80 and Poincaré's conventionalism; Lie's 21, Noether's 88
Theory   aim of (Poincaré) 187; Ajdukiewicz language as, 134, and meaning of scientific terms 192, and measurements (Duhem) 104, and observables 43, 93, 95, 104, 119–20, 126; commensurable vs. incommensurable 95; conceptual disparity between Lorentz's and Einstein's t. (Schlick) 164; content of (Poincaré) 63, 82–3, 88; constructive *vs.* principle t. 88, 170; deterministic t. and physics of central forces 78–9; empirical content of 42, 49–50, 59, 78; ephemeral nature of (Poincaré) 84, 116, 174, 187; *vs.* fact 4–5, 119; form of t. as its group and invariants x, 83, 189, 195; mechanistic, field and phenomenological t. 88; model–theoretic reconstruction 43; observationally equivalent (Lorentz's and Einstein's electrodynamics) 162–3, 168–9; pan-theoreticism 4, 119–20, 126; parallel (Poincaré) 181–2, 184, 193; principle theories *vs.* constructive theories (Einstein) 88, 170; relativity t. 151, 163, as a principles t. 170–1; invariants t. of a group (Klein) 166; relationship between Lorentz's Poincaré's and Einstein's t. of relativity 173; structure of 5, 42, 48, 51, 78, 81, 84, 88; set–theoretic reconstruction of 43, 107

view of conventionalist 181–4; holistic 50, instrumentalist 43, 52, 90–106, naïve-realist 174–6, 188, pragmatist 81, 83, Ramsey 42–3, 47–50, 81, 84, 88, 95, 164, 166, realist 43, 52, 82, 93, 95, 106, 169

theoretical terms as uninterpreted 60, as metaphors 31; elimination of, by Ramsey-sentence 42–3; theoretical neutrality of the physics of the principles 80, 84.

Time absolute (Lorentz) 154, 178, space-time 72, and Lorentz transformations 154, true, in Lorentz's (1904) 157, 168; *vs.* local time 157; fourth extra-spatial dimension (Hamilton) 85; conventions in the measurement of t. (Poincaré) 3, simultaneity of distant events (Poincaré) 178.

Transformation(s) Lie's theory of 21, 23, 29, 41, 65, 76; theory of dynamics 46, 73–4, 76, 77–8; contact (canonical) 22, 59, 65, 71, 74–5, 78, one-sorted and many-sorted 78; representation of advancing wave-fronts 65; construction of Huyghens's 59; generated by Hamiltonian function 72; continuous group of 29, 41; Darboux 84, 178; Galilean 169; Lie's theory of t. groups, variational calculus and Noether's theorem 88; linear, group of 87–8; Lorentz 1904, corrected by Poincaré in 1905/6, 156–7, 159, as rotations in a four-space (Poincaré) 158, approximate discovered by Larmor 190.

Translation dictionary for 8–9, 23, 34, 40, 84, 122; and invariants 4, 123, 139; between crude and scientific facts as indeterminate (Duhem) 125; Poincaré–LeRoy polemic 125–6, 138

   intertranslatability 8, 21, 25, 31, 37; of language 29, 31, 33, 59, 123, 176; of metric geometry 8–9, 22; between the wave and particle theories of light within geometrical optics 59

   non-intertranslatable languages 36, 40, 95, 137–9

True observationally 102, 107–8; "in nature" 102; indeterminate statements 102, 104–6, 116; no t. theory in the limit (LeRoy) 122;

   truth(s) absolute 105, analytic 116; relativisation of, to the conceptual apparatus (Ajdukiewicz) 133; in semantical sense and semantical paradoxes 140 and realist epistemology 141; "new theory of scientific truth" (LeRoy) 119

Vector(s) Fresnel's and Neumann's 183, 194, free of hypotheses 183; theoretical interpretations of 184; four-vector formalism and Poincaré xi, 72, 189; addition 66, multiplication 66.

World Euclidean and non Euclidean 126–7; and translation (Poincaré) 127; world-perspective 110, 128, 133, 139 and conceptual apparatus 130–4, 139 (Ajdukiewicz) world-picture 133; Lorentz's electromagnetic w-view 168, 170, 186 and Poincaré's epistemology 175–6